Modern Physics in 15 Minutes

Johan Hansson

Copyright © 2014 Johan Hansson

All rights reserved.

ISBN: 1479298050
ISBN-13: 978-1479298051

DEDICATION

This book is dedicated to Physics;
The quest to understand nature
at its most fundamental level.

CONTENTS

	Acknowledgments	i
1	Quantum Physics in 15 Minutes	3
2	Relativity in 15 Minutes	11
3	Particle Physics in 15 Minutes	21
4	Cosmology in 15 Minutes	31
5	The Big Bang in 15 Minutes	43
6	Black Holes in 15 Minutes	53
7	Chaos Theory in 15 Minutes	61
8	String Theory in 15 Minutes	73
9	Quantum Gravity in 15 Minutes	81
10	Bonus: Preons & Preon Stars	91

ACKNOWLEDGMENTS

To Richard Feynman and Fritz Zwicky;
my only true physics heroes.

1 QUANTUM PHYSICS IN 15 MINUTES

The physics we normally use to explain common phenomena is called "classical" because it is old and existed before the advent of the more modern quantum physics in the beginning of the twentieth century.

Quantum physics introduces huge deviations from what we normally consider to be "common sense". A specific cause no longer gives a unique effect. The reason is not that physicists want to be extra "mysterious" or "new-age"; it is simply that classical physics is totally wrong when it comes to describing microscopic "things", and also when one wants to understand common phenomena from a fundamental perspective.

In the end of the nineteenth century the general consensus was that physics was virtually complete, meaning that everything was fully researched and known, apart from some "small clouds on the horizon". Consequently, the young Max Planck got the advice not to get involved with physics as he then would waste both his time and talents. In the year 1900 Planck discovered quantum physics and the rest is, as they say, history.

Today we know that quantum physics is totally indispensable for really understanding most phenomena in nature. Some researchers even consider the whole universe as a quantum system, they then talk of quantum cosmology, in other words the attempted unification of quantum physics and cosmology - as the study of the universe as a whole is called.

What Planck, initially quite reluctantly, was forced to introduce into physics was the idea that light can only be generated in tiny, tiny energy "packets", or quanta. This is also the origin of the name "quantum physics" (or "quantum mechanics" which means the same thing) and was introduced mathematically through the introduction of "Planck's constant", h. The reason why this necessity had remained unnoticed was because quantum effects usually are unobservable on our normal, classical level. This is related to the fact that h is so very tiny that we do not notice the fundamental "graininess" of nature.

What is perhaps the oddest thing about quantum physics is that it is completely natural for a quantum mechanical object, for example an electron, to be at several places at once. We know, by direct experience, that a human being, a space shuttle or a mother-in-law (fortunately) cannot be at more than one place at a time. For an electron, a photon (particle of light) or any other quantum particle it is, on the contrary, more rule than exception. One and the same particle is everywhere simultaneously (in principle everywhere in the universe, but with different probabilities for different places) until it is actually observed by a measuring apparatus. Then it is "forced" to be at only one place. If one would distribute many measuring devices at several places, one can, in advance, predict the probability that the particle will be registered in any one of them. In principle, this is the only interesting thing, because it is anyway the only thing that one *can* calculate in quantum physics, i.e. the probabilities for different events to happen. But only sometimes is the actual position the interesting thing to know. Often one instead calculates the energy, the motion, transitions within atoms or atomic nuclei, etc., but *always* expressed as probabilities, never as certain, guaranteed outcomes.

One can interpret this as if nature *itself* does not have complete knowledge of exactly what happens that the world simply is "fuzzy" and uncertain at its most fundamental level. The real world truly is more fantastic than fiction. A side effect is also that everything which is not completely forbidden is compulsory.

For relatively low energies, quantum physics is described by the Schrödinger equation. With its help one can calculate probabilities for different physical processes. One can, at least in principle, calculate why different objects look, smell, taste, feel and react as they do. It is also possible to explain why and how chemistry and biology work, why there are different elements at all (i.e. to explain the structure of the periodic system) and why they have their different properties, how DNA works, how to design better "smart" drugs, electronic components, photovoltaic solar cells, and on and on. The list could go on forever.

In addition, the next big revolution in science, completely comparable to the industrial revolution and the information revolution (which we still are living in), will be the "quantum revolution". Computers that will defy today's imagination will be constructed on an atomic level making them unimaginably small and powerful, and they will be everywhere, both inside and outside our bodies, without even being noticed. It will be possible to store and transfer enormous amounts of information. Already it is possible to "teleport" photons, electrons and atoms - to instantly transfer them from one place to another without them having to travel through the space in between - and in principle no size limit for what can be teleported seems to exist. Maybe even the human teleporter of Star Trek can become a reality in a distant future?

To be crasser the big money will in the twenty-first century, and onwards, be generated by quantum physics, just like they today are connected to information (Maxwell's nineteenth century physics) and in the past were generated by steam engines and industries (Newton's seventeenth century physics).

At high energies the Schrödinger equation no longer correctly predicts the observed behavior, because it does not take relativity into account. A more correct version is called the Dirac equation. There are for example small, but completely measurable differences between the energy levels in an atom that are unexplained by Schrödinger but fully accounted for by the Dirac equation. Dirac also predicted, through his equation, the existence of anti-matter which shortly thereafter was discovered in experiments.

For extremely high energies particle-antiparticle pairs can be produced from pure energy. This is perfectly allowed according to Einstein's famous formula $E = mc^2$, which says that mass and energy really are two sides of a coin. Mass is just energy that has been "frozen in" and condensed into a particle. When lots of new particles are created/annihilated in for example a particle collider (where the necessary energy comes from the extremely high velocity, close to the speed of light, which the original colliding particles attain when accelerated) even the Dirac equation must be replaced by something called quantum field theory. Quantum field theory is the most exact language, so far, to describe nature. One may say that it mimics the most fundamental "grammar" of nature itself.

When also gravity – general relativity – is to be combined with quantum physics things go awry. (Quantum field theory only combines *special* relativity and quantum physics.)
There still does not exist a working and generally accepted theory for quantum gravity, despite what superstring theorists sometimes might want to lead you to believe. Simply stated, it means that we still do not have any microscopic theory, or "explanation", for gravity that is also experimentally tested and verified. Gravity is the only force that still just can be described in classical terms, without taking into account that nature on its most fundamental level seems to obey the strange rules of quantum physics. It is possible that the solution to this problem must await a new genius of Einstein's caliber.

Important people

Max Planck – Introduced the concept of the quantum for the very first time, in 1900, to solve "the ultraviolet catastrophe". Classical physics predicts that an object with a temperature above absolute zero should radiate an infinite amount of energy, which of course is in direct conflict with experience. Due to the fact that the atoms in an object, as Planck suggested, only can vibrate in given quantized configurations it becomes more and more difficult for the object to emit electromagnetic radiation with higher and higher energies (ultraviolet) because the energy-packets carried by each photon then becomes greater. In that way Planck solved this long-standing problem, and his formula completely explains the observed radiation. For this he got the 1918 Nobel Prize in physics.

Albert Einstein – Even though Einstein in his older days was one of the fiercest critics of quantum physics with famous remarks like "God does not play dice with the universe!", in his younger days he was one of the earliest and most important pioneers during the development of the early quantum physics. For example did he develop the theory that later made the laser possible (photons "like" to be in the same state as each other, they obey "Bose-Einstein statistics") and also deduced the ability different materials have for absorbing and storing heat from quantum mechanical calculations, something unknown before Einstein. He also, in 1905, explained the photoelectric effect, the ability ultraviolet light has to kick out electrons from metals, by assuming that light consists of a stream of particles; photons. The latter result won him the Nobel Prize in 1921. (Not the theory of relativity, as is commonly believed.)

Niels Bohr – Developed the first quantum mechanical model for the atom in 1913. Since the 1800s, physicists had observed unexplainable emissions of light, so called "spectral lines", coming from heated gases of different elements. Bohr explained why the spectral lines of hydrogen only occurred in specific colors by assuming that the electron in a hydrogen atom only could be located in specific shells around the nucleus. When the electron falls from a higher to a lower shell, the energy difference is radiated out in the form of a photon, explaining the different colors (also invisible "colors" in infra-red and ultra-violet that the human eye is insensitive to). Nobel Prize 1922.

Louis de Broglie – Bohr's model for the atom was, despite its success, very primitive and really based on old classical physics. For example, it could only be used for Hydrogen. Furthermore, Bohr could not explain why the different allowed "shells" existed, or why the electron was prohibited to fall all the way into the atomic nucleus, or in other words why there existed an innermost shell in the first place. Louis de Broglie was inspired by Einstein who had shown that light sometimes could behave as particles; even though classical physics says that light must be a wave. For nature to be consistent, de Broglie reasoned, matter, normally regarded to be particles, must under some circumstances be able to behave like waves. He therefore introduced his "de Broglie wavelength" that can be used for light and matter particles alike. De Broglie's wave-hypothesis was quickly verified experimentally by Davisson and Germer. Nobel Prize 1929.

Werner Heisenberg – Developed the very first detailed, logically consistent theory for quantum mechanics. Heisenberg's so-called "matrix mechanics" was based on the philosophy that only measurable quantities should be allowed in a theory. One problem was that most physicists at the time did not know the mathematics necessary for matrix mechanics, which made practical applications cumbersome and difficult. Nobel Prize 1932.

Erwin Schrödinger – Extended de Broglie's hypothesis of matter waves into a consistent theory for the dynamics of quantum mechanics in 1926. As the methods Schrödinger used were well known to all physicists (differential equations) the applications of the Schrödinger equation literally exploded in a large range of areas in atomic- and nuclear physics. Nobel Prize

1933 (shared with Dirac).

Paul Dirac – Developed the most general, and abstract, formulation of quantum mechanics, "transformation theory" in 1926. Dirac also showed that both Heisenberg's and Schrödinger's formulations were special cases of his own theory. He also developed the first relativistically correct quantum theory for matter in 1928. Nobel Prize 1933 (shared with Schrödinger).

Max Born – Heisenberg's supervisor during the time he, as a Ph.D.-student, developed matrix mechanics. Born also formulated the interpretation of quantum mechanics that still is considered to be standard. The wave function in Schrödinger's equation is not, as Schrödinger first erroneously believed, a description of how the electron in the atom is "smeared out". Instead, the wave function squared gives the probability that the electron, at a given measurement, will be found at a certain location. It is this interpretation that introduces the inherent uncertainty into quantum mechanics. Even if one is able to solve the Schrödinger equation exactly, the solution (the wave function) only contains information about statistical probabilities. Nobel Prize 1954.

Basic Equations

Heisenberg's matrix mechanics:

$$(XP)_{mn} = \sum_{k=0}^{\infty} X_{mk} P_{kn}$$

The Schrödinger equation:

$$[-\frac{\hbar^2}{2m}\nabla^2 + V(\mathbf{r})]\psi(\mathbf{r},t) = i\hbar \frac{\partial \psi}{\partial t}(\mathbf{r},t)$$

The Dirac Equation:

$$i\hbar \gamma^{\mu} \partial_{\mu} \psi - mc\psi = 0$$

MODERN PHYSICS IN 15 MINUTES

2 RELATIVITY IN 15 MINUTES

"Pure logical thinking cannot yield us any knowledge of the empirical world; all knowledge of reality starts from experience and ends in it"

Albert Einstein

Everyone has heard of Einstein and his theory of relativity. Most people are content merely knowing it exists, and some may even have heard about the formula $E=mc^2$.

One reason that few people have any deeper knowledge about relativity is that it, from the outset, got a reputation to be "impossible" to understand, or that only great geniuses could do so. That is not the case. Einstein's theories are today often encountered already in high-school, and in college relativity is commonly taught. Still, many of the consequences of relativity are in stark contrast to common sense.

Einstein constructed the theory of relativity in two steps. First, of course, the simplest: Special relativity (in 1905), which describes how things behave when moving at extremely high speed. In principle the theory of relativity is always more exact than Newton's laws which it replaces. But in practice, no real difference can be seen until the speed approaches about ten per cent of the speed of light (roughly 30,000 kilometres per second).

The special theory of relativity is based on two assumptions, or "postulates":

- The speed of light is equal for all, regardless of motion.
- The laws of physics are equal for all, regardless of motion (as long as the velocity is constant).

Even though these assumptions seem innocent enough, their consequences revolutionized physics.

But contrary to the common saying, everything is *not* relative. As we see, already in the assumptions themselves it is stated that the speed of light never changes – it is not relative but the opposite; absolute.

Everyone, regardless of their own motion, observe the same value for the speed of light. In relativity one can also construct other entities that never change their values. These unchanging objects are therefore called "invariants" and they can, if intelligently applied, simplify many of the calculations in the theory and also lessen the risk for misinterpreting the predictions, otherwise a common source of error in relativity. Actually, Einstein himself was on the verge of naming his new theory "the theory of invariance".

When the theory of special relativity was completed, Einstein could start his work on a more all-encompassing theory that was to apply more generally, not only for constant velocities. Consequently, the theory was named the theory of "*general* relativity".

This extension from special to general was no easy task. It took Einstein ten years to complete, mainly because he, at the outset, did not know the mathematics needed (differential geometry).

In 1915 the theory of general relativity was complete, i.e. the theory for general (also accelerated) motion.

As such, it is still our most exact theory for gravitation. The difference between Newton's old gravitational law and Einstein's general relativity could hardly be larger. They are simply two opposing ways of looking at gravity, so Einstein did not merely extend and expand Newton's framework but rather discarded it and started from scratch.

In general relativity there are *no* gravitational forces. Instead, a lump of matter (or energy, as it turns out to be the same thing from a gravitational viewpoint) distorts space-time, the four-dimensional amalgamation of the three dimensions of space and that of time. Objects then move "as straight as they possibly can" through this curved space-time (in technical language they are said to follow "geodetic lines", the equivalent of straight lines in curved geometries).

The result is that, for an observer, it seems like gravitational forces are at play influencing different bodies. One example is the planets in the solar system which travel "as straight as they can", as they are in free-fall. But because the sun's mass curves the space-time the end result is that planets go round and round the sun. Another example are stars in a galaxy which also are falling freely in space, but which in turn curves due to the mass in the galaxy itself, making the stars orbit around the galaxy centre.

For weak gravitational fields, i.e. for small curvatures of space-time, it is nearly impossible to measure the difference between Newton's and Einstein's predictions. In the solar system only very precise measurements can distinguish between the viewpoints, and this tiny difference is always in accord with Einstein, not with Newton. But on the other hand – if Newton's theory would have been completely wrong for the solar system it would of course not have been developed in the first place. To put a man on the moon in 1969 Einstein's theory was unnecessary – Newton's theory was completely sufficient for the precision needed.

For really intense gravitational fields in the universe, around very massive and/or compact objects, Newton's theory is not applicable and general relativity has to be used.

An extreme example is black holes which do not exist at all in Newtonian theory, while they automatically arise in Einstein's when sufficient mass/energy is concentrated within a certain volume. Theoretically one can even construct a black hole out of light alone. Even though the "constituents" are pure light the end result becomes completely black...

Perhaps the most complicated aspect of relativity is that time can pass at different rates depending on who measures or lives it. Time is thus "mouldable" and objects can really age differently, as it is time itself that "flows" differently, not only the clocks that measure it. And a biological life-form is really nothing but a (different kind of) clock.

If person A, at a distance, sees person B falling towards a black hole, then person A will never see person B entering the black hole. The miserable B will instead appear to freeze at the "surface" of the black hole, its event horizon. For person B herself it will however just take a brief moment before she is crushed at the centre of the black hole. This apparent contradiction is due to the fact that time itself behaves differently for person A and B.

In relativity both space and time are to a certain extent deformable. And even worse – time for one person can be space for another, and vice versa. The reason we find this weird, and maybe even unbelievable, is because we do not have direct experiences of these effects. What we call "common knowledge" or "intuition" simply is a result of what we, or somebody else, have experienced directly.

Even though we have no direct experience of the extreme

circumstances required to show relativistic effects, we still know they are correct. This is because we can perform experiments where the extreme conditions are realized. In all particle accelerators, which have been around in large scales since the 1950s, one has tested special relativity daily, among other things the world's most famous formula $E = mc^2$ which says that energy (E) and mass (m) are the same and can (almost) freely be converted into each other. A small amount of matter contains an enormous amount of "frozen" energy, as the mass m is multiplied by $c^2 = 90$ million billion (when m is measured in kilograms). This is utilized in, for instance, nuclear reactors – the radioactive waste weighs a tiny amount less than the uranium fuel cells. The difference is released as energy, out of which a small part is then extracted as electric energy.

It is thus known that special relativity is valid; while Newton's old theory for motion simply is wrong at these extreme circumstances. Newton did not have access to a particle accelerator, so we cannot really blame him for his theory's shortcomings. Additionally, his theory is valid to extremely high accuracy under "normal" conditions and energies.

General relativity has also been tested thoroughly, especially during the last thirty to forty years. Amongst other things, the physicists Pound & Rebka early on measured that time really do run slower at the bottom compared to the top of a high tower, exactly in accordance with the theoretical prediction from general relativity. Gravity, i.e., the curvature of space-time, is a bit stronger at the bottom of the tower. Again Newton is wrong; his theory predicts no difference whatsoever in the flow of time.

So far, there is only one practical application in ordinary life where general relativity has to be taken into account. Anyone who has navigated a boat in thick fog using GPS (global positioning system) would never have succeeded without the theory of relativity. Within the theory of relativity it is well known that time and space are not absolute, but rather depend on how something is moving (special relativity) and on the strength of the gravitational field (general relativity). These effects have also been tested in numerous experiments, and we know them to be correct. One such test involves sending a very precise clock on a journey on a plane, and then comparing it to another identical, initially synchronized clock remaining on the ground. When they are compared after the journey, they are no longer synchronized, and the difference can be completely accounted for by relativistic effects. And how does this connect to GPS? The GPS is based on a grid of satellites to make it possible, with high precision, to determine where on the Earth's surface one is situated. Relativity, or rather nature itself, however introduces two complications:

- The satellites are moving fast around the Earth. Even though the speed still is slow compared to the speed of light, it is enough to destroy the precision required for the positioning.
- Gravity is stronger at the surface of the Earth than it is up at the satellite. According to general relativity time flows faster for the satellite than for us on the surface. Again, the difference is minute, but sufficient to destroy the sought for precision. Only by correcting for the relativistic effects is it possible to attain the positioning accuracy needed.

The precision with relativistic corrections, where roughly 85 per cent comes from general relativity, the rest from special relativity, is about 15 meters. The "fuzziness" in position without consideration of relativistic effects would instead be of the order of kilometres, which would make the GPS-method useless for many practical purposes.

Important People

Albert Einstein - By the introduction of special relativity (1905) he solved a long standing enigma within physics; how to reconcile Newton's equations of motion with the more "modern" electrodynamic equations of Maxwell. Doing so, he also showed that the "ether", the conjectured medium for electromagnetic wave propagation, does not exist. Or at least is unnecessary for the theory.

Einstein again revolutionized physics with the general theory of relativity (in 1915) where he started from the simple but profound insight that for a person in free fall, for instance in an elevator where the cable has just broken, there is no gravitational force at all. In other words, it is only when non-gravitational forces keep a body from attaining its natural motion (free fall) that one experiences gravity.

Albert Michelson - Tried, for a long time, and through a series of increasingly more sophisticated experiments (together with Edward Morley) to measure Earth's motion through the ether. Very reluctantly, Michelson had to acknowledge that the ether actually doesn't seem to exist, a "non-result" for which he later received the Nobel Prize 1907.

Hendrik Anton Lorentz - Was the first to write down the relativistically correct transformations between two observers moving fast relative each other, which solved the incompatibility between Newton's equations of motion and Maxwell's electrodynamics. Lorentz's view was, however, that the ether mechanically compressed objects in the direction of their motion and did not reach the more elegant solution given by Einstein in his special theory of relativity (1905).

Hermann Minkowski - Einstein's mathematics teacher and the one who first saw that time should be treated as an extra, fourth dimension. Time is in relativity treated (almost) as a space-dimension, but in which one can move only in one direction, forwards.

Arthur Eddington - British astronomer who planned and carried out an expedition (1919) to measure the general relativistic prediction for how light should be bent due to the mass of the sun. As one used the apparent shift in the positions of the fixed stars, close to the edge of the sun, one had to await a total solar eclipse, giving a delay of four years after the theory was first proposed. The result of this expedition made newspapers across the globe launch Einstein as a genius, a hero and a world celebrity.

Basic equations

Lorentz transformations "translate" measurements and observations from one frame of reference to another, moving with velocity v relative to each other (along the x-axis in this case).

$$t' = \gamma(t - \frac{vx}{c^2})$$

$$x' = \gamma(x - vt)$$
$$y' = y$$
$$z' = z$$
$$\gamma = \frac{1}{\sqrt{1 - v^2/c^2}}$$

Einstein's general relativistic field equations constitute a coupled system of ten non-linear partial differential equations, describing the gravitational field, i.e., how mass and energy curves space-time.

They can be solved exactly only for a few simplified special cases.

$$R_{\mu\nu} - \frac{1}{2} R g_{\mu\nu} = \frac{8\pi G}{c^4} T_{\mu\nu}$$

The Schwarzschild metric is perhaps the most well-known solution to Einstein's field equations. It describes the "metric", i.e., the space-time geometry, around a spherical and static mass, M.

$$ds^2 = (1 - \frac{r_s}{r}) c^2 dt^2 - \frac{dr^2}{1 - \frac{r_s}{r}} - r^2 (d\theta^2 + \sin^2\theta d\phi^2)$$

The formula for a (non-rotating) black hole: If the mass (M) lies within a sufficiently small volume, it will collapse to a black hole. Nothing that enters inside the so-called "Schwarzschild radius", r_s, can ever get out, not even light (hence the name). One can view it as if space itself inside r_s is falling towards the centre faster than the speed of light. For Earth the Schwarzschild radius is of the order of one centimetre, but there is no known force that could compress the Earth that much.

$$r_s = \frac{2GM}{c^2}$$

The Friedmann-Robertson-Walker-metric is the basis for modern cosmology. If one idealizes the energy and matter content of the universe as being exactly evenly distributed in space (homogeneous) and the same in all directions (isotropic), the FRW-metric is the unique solution to Einstein's field equations.

$$ds^2 = dt^2 - a(t)^2 (\frac{dr^2}{1-kr^2} + r^2 d\theta^2 + r^2 \sin^2\theta d\phi^2)$$

Here *a(t)* is the "scale factor" of the universe, which in the FRW-model contains *all* the dynamical information of the theory.

3 PARTICLE PHYSICS IN 15 MINUTES

What is fundamental?

How many times is it possible to divide something? If we want to find the fundamental constituents of matter we must successfully answer this question.

Somewhere the limit probably will be reached and the smallest and most fundamental particles will then have been found. It used to be the atom, but today several much smaller constituents share the concept "indivisible". The model that presently best describes the innermost workings of matter is called the standard model of particle physics, and is described by something called quantum field theory.

Matter's most fundamental building blocks (known today) can be divided into two main types; "quarks" and "leptons". Quarks make up, among other things, protons and neutrons which in turn make up the atomic nuclei. An example of a lepton is the familiar electron, which conducts electricity in power lines, and also keeps human bodies and all other objects from disintegrating.

But not only matter, quarks and leptons, are particles; even the forces themselves are particles at this fundamental level. This is a consequence of the usage of quantum field theory to describe the forces. The particles simply are quanta, energy-packets of certain sizes, of the corresponding force fields.

We know of only four distinct natural forces – gravity,

electromagnetism, strong nuclear force and weak nuclear force. The latter three being described as an interchange of "force particles":

- Electromagnetic forces are mediated by photons (particles of light), of which there is only one kind.
- The weak force, among other things causing radioactive decays, is mediated by two charged W-particles and a neutral Z-particle.
- The strong force, that binds quarks into so-called hadrons, like the proton and the neutron, is described by eight different gluons ("glue particles"). A leakage of this gluon-force gives rise to the nuclear force binding protons and neutrons into different atomic nuclei.

Both nuclear forces are extremely short-ranged; they are active only inside the atomic nucleus, their "reach" a miniscule thousand billionth of a millimetre or less. Because of this we do not experience them directly in our daily lives and that is why it took until the 1970s to really understand them.

Interactions

One of the most important aspects of the matter particles is how they interact with each other; that is, how they are acted upon by the force particles. Quarks interact by all three types of forces. Electrons, and the other charged leptons, are acted upon by photons and W/Z-particles. Neutrinos, the electrically neutral leptons, only interact by means of W- and Z-particles – the weak interaction, having the shortest range of all the known forces. Therefore, neutrinos are almost completely free to fly through the cosmos; they rarely take part in any interactions whatsoever. The earth, and ourselves, are constantly being penetrated by enormous numbers of neutrinos, many of them made as intermediate by-products in the energy production inside the sun. All materials are, for all practical purposes, completely transparent to neutrinos.

Particle families

A peculiar and unexpected discovery during the 1960s and 1970s was that quarks and leptons come in "families" or "generations". So far one has detected three families, where quarks and leptons are arranged in pairs with increasing mass.

The first family consists of the quarks u (up) and d (down) and the leptons e (electron) and its neutrino, ν_e. In the second family we have the quarks s (strange) and c (charm) and the leptons µ (muon) and its neutrino ν_μ. In the third, and so far the last, family we have the quarks b (bottom) and t (top) and the leptons τ (tau) and its neutrino ν_τ. If this repeating structure continues (and in that case how far) is unknown. By measurements of the decay characteristics of the Z-particle it is known that any eventual further generations at least cannot have light or massless neutrinos, which the three known generations all have.

However, one does not have the slightest clue as to why this repeating structure is built-in into nature. The particle generations seem to be almost completely identical, except for the increasing masses for each generation. Also, some "quantum numbers" are different.

Already in the 1940s, when the muon was first discovered, this problem surfaced. Why should the electron have an almost identical "cousin" roughly 200 times heavier?

In the 1970s the next electron-cousin was discovered, the tauon, again identical to the electron but with a mass 3,500 times larger. That did not make the puzzle less puzzling.

Few particles explain many phenomena

If we count the number of particles that today constitute the most fundamental level in the universe we get, on the matter-side, six quarks (all with three different colours, making eighteen) and six leptons, and on the force-side one photon, three W- and Z-particles and eight gluons.

This may seem complicated, but one should remember that this so-called standard model of particle physics describes all known phenomena in the universe – bar one. There is still no correct formulation of the gravitational force in terms of particles. Today one has no understanding of how gravity arises from a microscopic viewpoint, our best theory of gravity still being Einstein's general relativity from 1915.

General relativity is a purely classical theory, which means that it is formulated without regard to quantum physics.

The problem of unifying general relativity and quantum mechanics, and in that way construct a theory for quantum gravity, is the biggest fundamental challenge in theoretical physics.

It is also entirely possible that quarks, leptons and force particles, in turn, may consist of even smaller, even more fundamental constituents. Why we have not seen them may be due to that we have not been able to look deeply enough into matter yet.

All the phenomena that man (so far) has encountered can thus be explained by only four forces of nature. Whatever will happen in the future, this is an enormous accomplishment and a simplification unique of its kind.

Of the four forces, we have direct experience of only two – gravity and electromagnetism. The remaining two are, as previously stated, only present within atomic nuclei – the weak and strong nuclear forces. The reason why we readily experience gravity and electromagnetism is that they have infinite reach, while the nuclear forces quickly fall to zero outside the nucleus.

Connection between the largest and the smallest

Today we believe that the universe was created out of nothing roughly 14 billion years ago, from an initial state that (perhaps) was infinitely compressed. To explain how it was created and evolved in its early stages needs quantum physics. In the newly born universe the natural length scale was comparable to the sizes of the elementary particles. As a consequence, the laws of particle physics will have determined how the early universe evolved. There is thus a natural and unavoidable coupling between the very smallest and the very largest – that is between the fundamental elementary particles and the universe as a whole.

If the natural laws in the early universe had deviated by only a small fraction from those we know today the present universe would have looked completely different. Some even go several steps further and claim that the universe was "made to order" to create intelligent life. That hypothesis is known as the "anthropic principle", but only a minority of scientists believe it.

The early universe is, strangely enough, easier to describe theoretically than the present universe. This depends on the fact that the early universe was very hot and uniform. We think that the physics already known and tested in the lab is sufficient to describe the universe as early as a fraction of a billionth of a second after its birth.

This may sound like hubris, but if we compare it to something we know about, for instance ordinary water, we can start to realize how this can be. If we heat water to several thousand degrees the result is plasma; an ionized gas of water, more or less identical everywhere. And when something is the same everywhere it is easier to describe mathematically.

When the water is cooled it first turns to normal gas – steam – as the electrons can bind to the nuclei when the temperature cools below the ionization temperature. As the temperature goes down further the water becomes liquid, and finally it turns into a solid – ice. The water becomes increasingly more difficult to describe in detail as we lower the temperature, as more and more phenomena and structure can arise. Ice has, for example, (at least) eighteen different possible crystal structures and complicated bonds between the electrons in the water molecules.

The universe works in a similar fashion – when the universe is cooled by the expansion, more and more complicated structures can start forming. In the early universe it is assumed that every part is identical to any other, except for the very tiny fluctuations that act like "seeds" for the structures we see around us today. In today's universe different forces, mainly gravity, have produced virtually infinitely many detailed structures on all length scales imaginable.

And if it really is the case that the universe today is dominated by a mysterious dark energy, as the best present observations seem to point towards, the odds are great that also dark energy arises from some (unknown) microscopic phenomenon – again a connection between the largest and the smallest.

Loose ends

Even though the standard model has been very successful, many questions remain. One is the family structure – why do the electron and the two lightest quarks have "carbon copies", heavier cousins? They are not needed, as far as is known, to build up any of the objects we actually observe, but can only be created "artificially" and then decay very rapidly to the first family.

Another loose end is the hypothesis about the Higgs particle. In the standard model there is the problem of how to generate masses for the particles. In 1963 the Scottish physicist Peter Higgs proposed that there might exist a special quantum field that introduces masses, and the corresponding particle has been dubbed the Higgs particle. But still, some fifty years after its conception, it is hypothetical – something akin to it has recently been observed. To show that it really is the standard model Higgs particle will need much more work. And even if it turns out to be it still does not explain the origin of the different masses of the elementary particles, as often erroneously stated.

The total knowledge of fundamental research is perpetually increasing. At the same time new questions pop up for each one that is being answered, especially in subjects as strongly driven by experimental results as particle physics was during the 50s, 60s and 70s. In recent years the fruitful influence from unexplained experimental results has largely been absent, partly as the standard model has been so successful that it has become somewhat of a "victim of its own success".

There are great expectations that the Large Hadron Collider, LHC, at CERN in Geneva will provide answers about the nature of the Higgs particle, and additional new experimental clues and conundrums. Historically what usually has happened when a new and more powerful accelerator has been deployed is that something completely unexpected has been found, something that no one had predicted. Nature herself is usually much more inventive than the mere mortals trying to describe her. A research area that is strongly advancing works like a rapidly inflating bubble. The inner volume, the known, is of course growing, but at the same time its surface – the boundary to the unknown, the cutting edge of research – is also growing at an accelerated pace.

Important people

J.J. Thompson – Discovered the first elementary particle, the electron, in 1897. This can be seen as the starting point for particle physics. Nobel Prize: 1906.

Ernest Rutherford – Discovered the proton (1919). Rutherford, who regarded himself as a physicist (discovered the atomic nucleus in 1911), was not glad to be awarded the Nobel Prize in *chemistry*. Nobel Prize: (in chemistry, for certain types of radioactive decays) 1908.

Paul Dirac – Invented quantum field theory, the mathematical "language" used to describe particle physics (1930). Quantum field theory is a union of quantum physics and the special theory of relativity. Nobel Prize: 1933.

James Chadwick – Discovered the neutron (1932). Nobel Prize: 1935.

Carl Anderson – Discovered the positron, the first example of an antiparticle (1932). The positron is an anti-electron. Nobel Prize: 1936.

Enrico Fermi – Suggested the first successful theory for the weak nuclear force, among other things the cause of radioactive decays (1933). Nobel Prize: 1938.

Hideki Yukawa – Suggested the first detailed theory for the strong nuclear force, which holds the atomic nucleus together (1935). Nobel Prize: 1949.

Richard Feynman – Theoretical physicist who made great contributions to many different areas, but particularly to theoretical particle physics. In the end of the 1940s he showed how quantum electrodynamics should be modified to make practical calculations possible. In connection with this he introduced the "Feynman diagrams" which in a simple way visualizes particle reactions and structures the calculations. He also formulated (together with Gell-Mann) the first detailed theory for weak interactions (1958). Nobel Prize: 1965.

Murray Gell-Mann – Dominated large parts of particle theory during the 1950s and 1960s. Suggested quarks (u, d and s) in 1964 (which independently also was done by George Zweig under the name "aces"). Formulated the basis for quantum chromodynamics (1972), the theory of how quarks interact by means of "glue particles" (gluons), together with Harald Fritzsch. Nobel Prize: 1969.

Steven Weinberg – Weinberg, together with Gell-Mann, dominated particle theory during the 1960s and 1970s. He contributed greatly to all components of the present standard model in particle physics, which summarizes/explains all experimental data up to the present. Nobel Prize: 1979.

Sheldon Glashow – Planted the seed to the unified electroweak theory already in his Ph.D. thesis in the 1950s (which was written with the famous physicist Julian Schwinger as supervisor). Were also the first to propose a fourth quark, the charm quark (together with James Bjorken). Nobel Prize: 1979.

Basic equations

Quantum electrodynamics (QED):

$$L = \bar{\psi}(i\gamma^\mu D_\mu - m)\psi - \frac{1}{4}F_{\mu\nu}F^{\mu\nu}$$

Modern quantum field theories are usually formulated starting from a "Lagrangian", L, that roughly speaking describes the energy at every spacetime point. Ψ here describes charged matter fields (quarks and leptons), F describes force fields (photons) and D contains the term responsible for the exchanged force (interaction).

Quantum chromodynamics (QCD):

$$L = \bar{q}(i\gamma^\mu D_\mu - m)q - \frac{1}{4}G^a_{\mu\nu}G_a^{\mu\nu}$$

Despite the superficial resemblance to QED, QCD is actually substantially more complicated. This is due to the fact that the gluons (the force particles) also interact among themselves, which the photons do not. q = quark field, G = gluon field, and D describes the interaction.

4 COSMOLOGY IN 15 MINUTES

The theory for everything.

To take a collective grasp of the universe as a whole may sound like the ultimate hubris. But surprisingly enough, scientists have been able to figure out much that was not even dreamed of fifty years ago. But with new knowledge new questions arise, and from having "known" one hundred per cent of the constituents of the universe we now only know about four per cent – which, at first sight, may not look like progress at all.

Why isn't the sky as bright as the sun, 24/7? It is obvious, you may think? But the truth is that this used to be a big paradox until fairly recently, which went under the name "Olber's paradox".

Let us for simplicity assume that all stars are like the sun. The sun's intensity decreases with distance. If we imagine a spherical bubble with the sun at its centre the total sun-light spreads out on an increasingly larger area when the bubble, i.e. the distance, gets larger. The surface of the bubble increases as the distance squared, that is the distance times itself. This means that we get the formula $1/4\pi r^2$ for describing the light intensity at a specific distance (r).

So far this may be seen as an explanation for why it is dark at night. But there is another aspect of the formula when applied to the universe – the number of light sources (stars) increase with the distance from us. The farther away we look the more sources are sending their light towards us.

If we assume that the universe looks roughly the same everywhere (this assumption is called "homogeneity") then we know how many stars there are in a given volume.

If the stars, in the mean, are as numerous regardless of where in the universe we are then the number of stars on the bubble referred to above will increase with the area. If we increase the distance to our imaginary bubble so that its area is doubled the number of stars on the bubble surface also is doubled. The number of stars increases in exactly the same manner as the light intensity from an individual star decreases. The effects thus cancel each other.

If we also take into account that stars at different distances partly shadow each other on the line of sight the conclusion is that the whole sky should be as bright as the sun itself, both day and night. But this is of course not what we observe – and that's Olber's paradox.

The solution

What, then, is the modern resolution to Olber's paradox?

1. The Universe expands
2. Stars do not live forever

In the 1920s Edwin Hubble, using the world's largest telescope at the time, observed that the universe expands. He discovered "Hubble's law" which states that the rate of expansion is directly proportional to the distance. If a galaxy is at a specific distance from us and recedes from us at a specific velocity then another galaxy, at double the distance, will recede twice as fast. It is the easiest relation imaginable, and directly results if one assumes that the universe expands as a "muffin dough": the dough in-between the raisins (the voids between the galaxies) expands at the same rate everywhere, and as there is more dough between us and the raisins farther away they will automatically recede faster from us.

After thinking about it for a while one concludes that it doesn't matter what raisin we are situated on, all other raisins recede from any given one and with a higher velocity the farther their distance.

As the same thing seems to apply to galaxies in the expanding universe the conclusion is that there exists no centre in the universe. Wherever in the universe we are situated the expansion looks the same. This is called the *cosmological principle*: at really great distances the universe is the same everywhere (homogeneity) and in all directions (isotropic). One should however keep in mind that the cosmological principle is an *assumption* that seems to be fulfilled as far as we know based on present observations. There is nothing to forbid that the assumption may be proven wrong in the future.

The expansion has two effects on the starlight. First, the light is distributed on an area larger than the one we previously assumed – when the light finally reaches us the bubble has expanded in size due to the expansion. Secondly, the light itself is "stretched"; its wavelength expands in step with the universe. When the wavelength of light increases its energy decreases (red light with longer wavelength has less energy than blue light with shorter wavelength). The intensity of the cosmic light is diluted by these two effects. As we today also know that stars have a limited life-span (large luminous stars have much shorter lives than small and dim ones) they send out only a limited amount of light-energy during their lifetimes.

When the effects from the expansion and the lifetimes of stars are combined it suffices to explain why the sky is dark at night. The small specks of light from stars visible to the unaided eye all lie in our own galaxy, and furthermore very "nearby" in astronomical terms.

Astrophysics and Cosmology

Somewhat simplified one may say that astrophysics tries to explain astronomical phenomena by using known physical laws. (One hope is also to find brand new laws of nature.) Cosmology, on the other hand, is the science of the universe as a whole; its origin, development, and final fate. Within cosmology, and large parts of astrophysics, gravity is the dominant force.

Even though gravity is by far the weakest of the known forces it has a property that no other force has: it is always attractive. (One possible exception is "dark energy", see below.)

This means that gravitation always adds so that it eventually overpowers all other forces, but only in the presence of enough matter and/or energy. For all large-scale structures in the universe: solar systems, galaxies, galaxy clusters, etc., gravity completely dominates. To be precise gravity is no force at all, according to Einstein it is an effect of matter and energy curving space-time itself. That is why one today always talks about relativistic cosmology.

CMBR – The oldest photograph we can take (COBE/WMAP)

The universe is filled with left-over heat from the Big Bang. This is called the Cosmic Microwave Background Radiation (CMBR or CBR – cosmic background radiation). This background light had a temperature of 3000 Kelvin when the universe became transparent to light for the very first time, roughly 400 000 years after the Big Bang. Before that time the universe consisted of plasma, mostly ionized hydrogen gas, which light cannot travel through. When the universe, through its expansion, had cooled enough the plasma turned to neutral hydrogen gas and the universe became transparent.

The universe has since expanded roughly 1000 times in size, which also means that the background radiation temperature has decreased with a factor of 1000. The temperature of the universe today is thus three degrees above absolute zero.

As the universe was non-transparent to light before the background radiation was released it also is an imprint, or "fossil", of how the universe looked when it was 400 000 years young. In other words the CMBR is the oldest photograph we can take of the universe.

In 1964 this invisible background light was accidentally discovered by Arno Penzias and Robert Wilson when they were in the process of measuring something completely different. This illustrates a recurring phenomenon in science – the really big discoveries are almost always made by chance.

Using the satellite experiment COBE (COsmic Background Explorer), John Mather and George Smoot could, in the early 1990s, determine the temperature of the background radiation, i.e. the temperature of the universe, with great precision: The answer: 2.725 Kelvins.

Starting in 2003 the satellite WMAP (Wilkinson Microwave Anisotropy Probe) measured tiny, tiny differences in temperature in the background radiation, which correspond to the state of the universe at an age of 400 000 years. This data contain large amounts of information, for example about the total energy of the universe. In addition, the slightly hotter areas should correspond to the "seeds" of over-density that are believed to having spawned the "super clusters" of galaxies seen in the universe today; if there are over- and under-dense regions gravity and expansion does the rest, automatically diluting the under-dense regions while contracting the over-dense regions to make them the gravitationally bound systems we see in the universe.

The cosmic background radiation is also one of three clues to why we believe in a Big Bang in the first place – a birth from nothing roughly 14 billion years ago. The other two clues are the expansion of the universe and the amount of Helium in the universe.

Can we ever take older pictures of the Universe?

Today it is impossible to look back to the early infancy of the universe.

The cosmic microwave background radiation is the oldest possible normal photograph of the universe and shows how the universe looked 400 000 years after its birth.

But just like there is a left-over radiation in light (photons) theoretically there should also be "primordial" radiation consisting both of neutrinos and gravitational waves (which may be described by gravitons) which both should have been produced copiously in the Big Bang. Neutrinos and gravitons hardly interact with matter at all, which means that the universe became transparent for these two types of radiation much earlier than 400 000 years.

The drawback is that just because they are so reluctant to interact we have yet to detect any of these imprints of the Big Bang – they are simply unobservable with today's technology. Some scientists even believe that it never will be possible to detect these signals.

But one should be careful to declare something impossible. Wolfgang Pauli, who in 1930 invented neutrinos theoretically, stated that they would never be seen in experiments. But already in 1956 Frederick Reines and collaborators detected neutrinos experimentally using a nuclear reactor, something Pauli could not have foreseen in 1930.

If neutrinos from the Big Bang ever are detected this would mean that we get information about the universe at an age of only one second, as the universe became transparent to neutrinos after roughly one second.

Using gravitational waves from the Big Bang one could, in principle, get a "photograph" of the universe 10^{-30} seconds after the origin of time. The universe was then merely 0.000000000000000000000000000001 seconds old!

Dark matter

There are indirect indications that there exist huge amounts of "dark" invisible matter in the universe. So far there are three clues pointing to this conclusion:

- Spiral galaxies rotate too fast to be stabilized from disintegration by gravity from the visible matter alone. There must be large amounts of dark, unknown, "exotic" matter within them, helping them keep together. A typical spiral galaxy seems to contain roughly ten times as much dark matter than normal, visible matter. An additional enigma is that the other main type of galaxy (elliptical galaxies) does not seem to contain much dark matter at all, if any.
- The galaxies in galaxy clusters move too fast for the cluster to be stable. Again dark matter is needed to keep them together. The amount of dark matter needed for galaxy clusters seems to be even larger than for individual spiral galaxies. Independent measurements of the masses of galaxy clusters using "gravitational lensing" give roughly the same amount of dark matter as the one calculated from the cluster's dynamical motion.
- Measurements on the universe as a whole, for instance by using the microwave background radiation, and the appearance of enormous large-scale structures like cosmological "voids" and "filaments" – rope-like structures along which all known galaxies seem to lie – also point to the

conclusion that the amount of dark matter must be many times higher than visible, normal matter.

There are many candidates for dark matter, especially hypothetical particles from untested theories in particle physics. The problem is that these particles have never been seen in earth- bound experiments. So far, the only indications for dark matter come from indirect astrophysical and cosmological observations. But the hunt continues, as an understanding of matter and at least one, but probably a couple of, Nobel Prizes is at stake.

An alternative to dark matter that has been proposed is that gravity maybe behaves differently than we think at really large distances and/or low accelerations. In that case no dark matter is needed at all. The most well-known such theory is called MOdified Newtonian Dynamics, or "MOND".

The enigma dark energy

Dark energy is something completely different than dark matter. Astronomical measurements since 1998 seem to show that there is a repulsive gravitational force, effective only at the very largest, cosmological distances. It is noticeable only when the universe is studied as a whole.

In contrast to dark matter there are not even any hypothetical suggestions as to what dark energy may be. It is admittedly easy to insert a term representing dark energy in Einstein's general relativistic field equations and in that way model the behaviour one apparently sees. But there exists no physical explanation of the underlying "mechanism" or of what dark energy *is*. The most common term used is called the "cosmological constant", which simply expressed is a constant energy that exists in vacuum. What this energy consists of is unknown, but whatever it is it must have negative pressure which gives rise to repulsive gravitation when inserted into Einstein's equations. The problem is that all known matter and radiation give rise to positive pressure. Dark energy can thus not be explained by any known particle or radiation, or by any of those conjectured but not yet seen.

Visible, normal matter seems - according to the latest combined observations - to make up only four per cent of the total energy in the universe. 23 per cent seems to be exotic dark matter. The rest, 73 per cent, should then be some mysterious form of dark energy that we have no theoretical explanation for whatsoever.

In the 1970s the consensus view was that we knew what 100 per cent of the universe consisted of (normal matter and some radiation). Today this number has decreased to just a few per cent. If the trend continues it could mean that soon we will not know anything at all about the content of the universe...

This is, however, typical for a research field in such great expansion and evolution as cosmology. For every question answered at least ten new questions surfaces.

Important people

Albert Einstein – Formulates general relativity, our modern theory for gravitation (1915). Suggests a modifying addition to the theory, the cosmological constant (1917), to make a static, unchanging universe possible, as this was the accepted world-view at the time.

Georges Lemaitre – Suggests the original Big Bang-theory (1927), that the universe was created from a "primordial egg".

Edwin Hubble – Discovers other galaxies – our galaxy evidently is not the whole universe, as previously believed (1924). Hubble discovers that the universe is expanding (1929).

Alexander Friedmann, Howard Robertson & Arthur Walker – Formulates (1922-1935) the "standard model" of cosmology that is used to this day. It assumes that the cosmological principle applies to the universe as a whole.

Fritz Zwicky – Proposes dark matter to explain how clusters of galaxies can rotate so fast without disintegrating (1933).

George Gamow – Calculates that a universe which started in a Big Bang must result in a left-over background radiation "glow" with a temperature of 5 Kelvin (1948). This theoretical result is then promptly forgotten for more than fifteen years…

Arno Penzias & Robert Wilson – Discover the cosmic background radiation, a cooled remnant of the Big Bang (1965), for which they receive the Nobel Prize in 1978.

Alan Guth – Proposes an extremely fast, short expansion phase – inflation – in the very early universe (1980). This results in that a very small region incredibly fast is inflated into what becomes our whole observable universe. Inflation is nowadays most known for curing several problems in the original Big Bang-model, for instance why the (observable) universe should have the same temperature everywhere.

John Mather & George Smoot – In the early 1990s they measure the temperature of the universe with very high precision. The universe is seen to be a nearly perfect black-body radiator with a temperature of roughly 3 Kelvin. They also see (small) temperature fluctuations in the background radiation. For these discoveries they get the Nobel Prize in 2006.

Supernova Cosmology Project & High-z Supernova Search Team (1998) – Observe, completely unexpected, that the expansion rate of the universe seems to be increasing instead of decreasing as everybody previously believed (as gravitation should act like a brake). This suggests that mysterious dark energy dominates the present universe, which leads to a renaissance for Einstein's cosmological constant.

Basic equations

The Friedmann-Robertson-Walker (FRW)-metric: This is the unique solution to Einstein's equations under the assumption that the universe is the same everywhere (homogeneous) and in all directions (isotropic) on cosmological distances. This assumption is called the "cosmological principle". The FRW-metric describes the four-dimensional space-time geometry of the universe.

$$ds^2 = c^2 dt^2 - a(t)^2 \left(\frac{dr^2}{1-kr^2} + r^2 (d\theta^2 + \sin^2\theta \, d\phi^2) \right)$$

Friedmann's equations: Describe the dynamics of the idealized FRW-universe, which is contained solely in the cosmic scale factor, $a=a(t)$, defined in the metric above. $a(t)$ is a measure of how the emptiness in the universe expands. Friedmann's equations are extremely simplified special cases of Einstein's general relativistic field equations, applying solely under the assumption of the cosmological principle:

$$\left(\frac{\dot{a}}{a}\right)^2 = \frac{8\pi G}{3}\rho - k\frac{c^2}{a^2}$$

$$\frac{\ddot{a}}{a} = -\frac{4\pi G}{3}\left(\rho + 3\frac{p}{c^2}\right)$$

The cosmological red-shift parameter, z:

$$z = \frac{a(t = now)}{a(t = then)} - 1$$

Hubble's law:

$v = H_0 \, d$

Where H_0 = Hubble's constant, v = recessional velocity, d = distance

5 THE BIG BANG IN 15 MINUTES

The best astronomical observations of today suggest that the universe was created about fourteen billion years into the past. Only a few decades ago, most physicists saw with deep skepticism on statements such as these, and only recently have cosmology, the study of the universe as a whole, started to be regarded as legitimate research by the majority of researchers. The reason is that modern technology has made it possible to make detailed observations to test and refine the models of the universe, although many problems still remain. Today's best model is probably still a crude simplification of reality.

Since light is moving at a very high but finite speed, we can actually do a sort of "CAT-scan" of the universe - both in space *and* time. We see very distant objects as they were a long time ago, when they sent out their light that just about now is reaching us, as the universe was much younger when that light was emitted. (Kind of like subscribing to a newspaper from Argentina via cargo ship from South America, the "news" that has been traveling a long time is already history when it reaches us.) Hence it is possible to make an "archaeological" survey of the universe, not by digging deeper and deeper through layers of soil, but by looking farther and farther out into the universe.

The currently most successful model of the universe is called the "Big Bang". Unfortunately, the name is a bit misleading because one might believe there was an ordinary explosion in an already existing space. On the contrary, both space *and* time were created simultaneously with all matter and radiation. What existed before the Big Bang thus loses its meaning in that (strictly classical) model because time itself did not exist "before". A quantum mechanical theory of gravity might be able to ask and answer such questions as "what preceded, or caused, the Big Bang", but unfortunately such a theory does not yet exist...

Why we believe in the Big Bang?

There are three main clues that today point towards the conclusion that the universe previously (probably) was in a highly compressed state.

The most important clue, because it is difficult to explain in other models, is the leftover glow from the past hot, compressed, state of the universe. It was accidentally discovered in 1964 by Penzias and Wilson. However, Gamow had, already in the 1940s, calculated that a universe which begins in a Big Bang would produce a radiation temperature today of five degrees Kelvin, that is, five degrees above absolute zero. Penzias and Wilson was unaware of that prediction, and their discovery, which they later received the Nobel Prize for, began as just an annoying noise in their apparatus, which disturbed what they initially really wanted to measure. They saw a temperature of three degrees Kelvin, the same in all directions over the sky. This is precisely what one would expect for the residual glow from the Big Bang.

The second clue is the abundance of helium in the universe, which is far too high to simply having been produced in the stars (which, as their power source, convert hydrogen into helium during most of their lives). If one uses ordinary nuclear physics, the correct number easily and more or less automatically emerge out of a Big Bang model. During the universe's first few minutes these fusion processes produced one quarter helium from the initial hydrogen, then it became too cold, because of the expansion, for the mechanism to combine any more hydrogen nuclei (protons) into helium nuclei, and nuclear fusion stopped - until the first stars lit up several billions of years later.

The third and final clue is Hubble's discovery from the late 1920s which showed that the universe is expanding. If you "play the movie backwards" it means that the universe was more compressed farther back in time. As previously stated, the best combined cosmological measurements indicate that the universe today is about fourteen billion years old (assuming the Big Bang model is correct).

The history of the universe, as far as we know...

It may seem odd to worry about, and to distinguish between, extremely short periods of time in the early universe. What can really be the difference between what happened at one millionth of a second after the Big Bang and at a billionth of a second, and how can it possibly affect us at all today? Then one is forgetting several things: First, the natural distance in the early universe was comparable to the smallest known elementary particles, and therefore the evolution of the universe depended very strongly on the elementary particles and the forces active between them at that moment. A very tiny difference in these properties could potentially have an enormous impact on how the universe works and looks like today. Second, in the early universe each factor of ten in time (e.g. the difference between 10^{-10} and 10^{-11}) is about equally important. Although the time interval is ten times smaller for each step closer towards the origin, all reactions also take place about ten times faster because the physical universe was then much smaller, so roughly as much "important physics" can happen in each such interval (mathematically it is said that this so-called "logarithmic time" is the natural time scale to compare and understand these phenomena). If this really is true (which no one really knows today), there is no end to the possible physics one can encounter as time zero is approached. We can divide something up in ten pieces and ten again and again... countless of times before reaching zero. It can, on the other hand, also be that there is a minimum "unit of time", the so-called Planck time, which has a value of approximately 10^{-43} seconds and which can be created from the three fundamental constants h, c and G. In that case you could divide one second in ten 43 times in succession before reaching this limit.

Observable signals from the very early universe can also be used for indirect verification/falsification of new, exotic physics, a kind of "poor man's accelerator".

MODERN PHYSICS IN 15 MINUTES

The following is a short, very preliminary and partial, history of the universe's epochs and the particles and forces that were then operating. After the electroweak epoch has passed, no "exotic" untested physics is needed, and it is believed that we currently have a relatively good understanding of the later processes.

Planck epoch ($t < 10^{-43}$ sec)

Since we are still unable to unite quantum physics (our successful theory for the very small) and general relativity (our successful theory for gravity) we are today unable to describe this era. Possibly, all four presently known forces of nature combined into one in this epoch.

GUT era ($10^{-43} < t < 10^{-38}$ sec)

The universe may here have included only two forces, gravity, and a so-called GUT-force (GUT for "grand unified theory") in which the two nuclear-forces, the strong and weak, and the electromagnetic force all were united into one. When the universe was 10^{-38} seconds old it had, automatically by the expansion, cooled to 10^{29} degrees, and then the strong nuclear force, in theory, separated to end the GUT-phase. The energy released by this "freezing-out" may have caused a sudden and dramatic inflation in the size of the universe.

Electro-Weak epoch ($10^{-38} < t < 10^{-10}$ sec)

The universe here had three distinct forces: gravity, the strong nuclear and the electroweak forces. When the universe was 10^{-10} seconds old, the temperature had dropped to 10^{15} degrees, which resulted in the separation of the electromagnetic and weak nuclear forces. That this really happens was verified indirectly through the detection of W and Z particles in particle physics experiments in 1983.

Particle-era ($10^{-10} < t < 10^{-3}$ sec)

Today's four forces of nature were now distinct. Particles and photons were equally abundant. When the universe was 10^{-4} seconds' old quarks combined into protons, neutrons and their antiparticles. At 10^{-3} seconds the temperature of the universe was 10^{12} degrees. Protons, anti-protons, neutrons and anti-neutrons could now no longer be created from the radiation. The remaining particles and antiparticles annihilated each other and created photons. A very tiny excess of particles over antiparticles became what we now see as matter in the universe. There is still no real explanation for how and why this imbalance was created.

Nucleosynthesis epoch (10^{-3} sec < t < 3 min)

During this era protons and neutrons began to combine through nuclear fusion, but the new heavier nuclei were also torn apart by the high temperature. After three minutes the universe had cooled to 10^9 (one billion) degrees, and the fusion stopped. At the end of this epoch the mass of the universe were 75% hydrogen and 25% helium, with only minimal traces of other nuclei such as deuterium (hydrogen with one extra neutron) and lithium.

Nuclear-epoch (3 min < t < 400 000 years)

The matter in the universe was now a hot plasma of hydrogen and helium nuclei and free electrons. Photons continuously collided with electrons, so the universe was still opaque to light. When the universe had been expanding for 400 000 years, it had cooled to a temperature of 3000 Kelvin. The electrons could then, for the first time, bond to atomic nuclei. Because the resulting atoms are electrically neutral, photons could now freely traverse the universe, which became transparent to light for the first time ever. It is this glow - cooled a thousand times as the universe has since become a thousand times larger because of the expansion - which Penzias and Wilson discovered in 1964.

Atomic-era (400 000 < t < 10^9 years)

The universe was now filled by atomic gas. Since no stars had yet lit up, and no other light sources existed, this is usually referred to as the "dark ages" of the universe. Density variations in the gas, combined with the gravitational attraction (probably "supercharged" with dark, exotic matter) formed the first "proto-galactic" clouds. The very first stars then were born, and in turn triggered the formation of the first galaxies from these enormous gas clouds.

Galaxy-era ($t > 10^9$ years)

The first galaxies are believed to having been born about one billion years after the Big Bang. In particular, the Hubble Space Telescope has recently observed that they were smaller, more numerous and fainter than the galaxies that exist today. This is taken as evidence that the smaller structures formed first, and then merged into increasingly larger objects, including clusters of galaxies, and clusters of clusters of galaxies (the so-called super-clusters) and giant cosmic voids (seemingly empty regions) and filaments – enormous structures along which the galaxies are concentrated - on the surface of the "bubbles" as defined by the voids. These, the largest of all known structures in the universe, can be hundreds of millions of light years big, and recently a void with an extent of over a billion light years was observed. The galaxy-era is the current era of the universe.

Important people

Alexander Friedmann - Showed in the early 1920s that Einstein's general relativity has solutions corresponding to an expanding universe.

Edwin Hubble - Discovered in the late 1920s, through astronomical observations with the then most powerful telescopes that the universe is expanding.

Georges Lemaitre - proposed in the early 1930s that the universe was created from a cosmic "primordial egg". This is usually seen as the start of the current big bang model.

Hans Bethe - Presented by the end of the 1930s the theory for how the energy of stars is created from nuclear physics reactions. It was quickly realized that all the helium in the universe probably could not have been created in stars.

George Gamow – Developed, in the 1940s, Lemaitre's theory and expanded it to include the early nucleosynthesis in the universe, mainly the production of helium, based on Bethe's nuclear physics. Gamow also realized that such a Big Bang model automatically predicts a residual glow that today should have a temperature of a few degrees Kelvin.

Fred Hoyle - coined the name "Big Bang" in an attempt to discredit the idea, as Hoyle himself (along with Bondi and Gold) in the 1950s had created an alternative to the Big Bang model; the theory of the "stationary state" of the universe, where it was assumed having always existed and looked much like it does now. The name, however, was quickly adopted by the Big Bang model *supporters*.

Arno Penzias & Robert Wilson - Discovered in 1964, by sheer coincidence, the residual glow from the universe's early childhood, and measured it to be roughly 3 degrees Kelvin. Penzias and Wilson had never heard of Gamow's prediction, predating their discovery by more than fifteen years. Ironically, Robert Dicke and his research group had just constructed a dedicated instrument to try to detect Gamow's proposed radiation, but had to see themselves beaten by Penzias and Wilson's serendipitous discovery. The satellite experiments COBE (1990) and WMAP (2003) have since measured the background radiation characteristics with very high precision, and are now being followed by the Planck satellite mission.

Basic equations

The relative number, N, of neutrons (n) and protons (p) in a given volume of the early universe is given by

$$\frac{N_n}{N_p} \approx \exp\left[-\frac{(m_n - m_p)c^2}{k_B T}\right]$$

When the universe cooled, the protons could no longer be converted to neutrons (due to the neutron's slightly higher mass m), while neutrons continued to decay through normal radioactive beta decay. The end result of the nucleosynthesis in the early universe is that about ¼ of the mass of the universe (at least of normal matter) today is helium, the rest being protons (hydrogen).

The temperature T is inversely proportional to "the cosmic scale factor" $a(t)$ at any given age, t, of the universe (where $a(t)$ in turn is related to the size of the observable universe for any t).

$$T \propto \frac{1}{a(t)}$$

In the early Universe most of the energy was contained in radiation, the universe is said to have been "radiation dominated" and $a(t) \propto \sqrt{t}$. As the universe became transparent to light, after about 400 000 years, it instead became the "matter-dominated" universe we still live in, which gives $a(t) \propto t^{2/3}$. It was only then that gravity could start building structures in the universe, before that time the intense radiation destroyed all objects that gravity tried to merge.

MODERN PHYSICS IN 15 MINUTES

6 BLACK HOLES IN 15 MINUTES

Many mistakenly believe that a black hole acts like a giant cosmic vacuum cleaner, devouring everything in its vicinity and eventually everything in the universe. If you were unlucky enough to get near a black hole you would certainly get sucked into it, and quickly be torn apart and ultimately crushed in its center, but at greater distance a black hole acts just like any mass whatsoever. If the Sun magically could turn into a black hole (which it never will as the sun is just too light) the earth and all the other planets in the solar system would continue in their orbits around the newly formed black hole as if nothing had happened. The only difference would be that the sunlight disappeared.

Black holes have a long and colorful history. A prerequisite for the black hole is general relativity - still our best theory of gravity, because black holes cannot exist according to Newton's law of gravitation from the 1600s (when the speed of light was assumed to be infinite). Einstein published his general relativistic field equations of gravitation in 1915. In general relativity, there are no gravitational forces at all; what we perceive as a gravitational force is in fact an effect of space and time - space-time - being curved by energy, or mass, since according to relativity theory mass and energy are two expressions for the same thing.

Already the following year, in 1916, Karl Schwarzschild found a solution to Einstein's field equations for a spherical, static, non-rotating mass. This means that space-time around, for example, the sun can be described well by Schwarzschild's solution. The reason it works in the solar system is that the sun is almost spherical, rotating slowly and loses mass (because of its energy production) very slowly. In addition, the sun contains more than 99% of the total mass in solar system, which means that as a first approximation it is perfectly alright to neglect the curvature generated by the planets (where Jupiter dominates as it is heaviest).

For "normal" astronomical objects the Schwarzschild solution does not create any problems. But if you somehow could compress all the mass inside a certain critical volume, i.e. within a given radius - the Schwarzschild radius - it was quickly realized that a lot of strange things would happen to Schwarzschild's solution. Singularities, where the density and consequently space-time curvature becomes infinite, would show up. One way to tackle this was to deny that matter could ever be compressed that much, the standpoint of Einstein himself. He even published an article in which he "proved" that it was impossible.

However, in 1939 Robert Oppenheimer and Hartland Snyder showed that a star with mass greater than a few solar masses automatically and inevitably will collapse to Schwarzschild's "impossible" objects. As Oppenheimer and Snyder had to idealize the problem to be able to solve it - among other things, it was assumed that the matter inside did not exert any pressure - most physicists were still skeptical.

Today we know for certain that general relativity predicts that these "exotic" objects exist as a possibility, from exact "singularity theorems" by Roger Penrose, derived directly from the theory of relativity itself, and in 1967 they were christened "black holes" by the American physicist John Wheeler. Why "black"? Because everything that ventures inside the so-called "event horizon" - which in the case of a spherical static mass is a fictitious spherical "surface" situated at the Schwarzschild radius - will be drawn into the center. This applies to everything, even light, which means that the black hole is... black.

Another way to put it would be that, as seen from the outside, time will slow down more and more for an object approaching the event horizon which means that light trying to escape changes its color towards lower and lower energy. Precisely at the event horizon this gravitational "red-shift" is infinitely large and the energy of the light zero, which of course means that there is no light. An object, for instance an astronaut, which falls towards the event horizon, experiences no such shift however. It crosses the event horizon without any problem and in a finite (and usually very short) time falls into the singularity and is compressed out of our universe. This apparent "paradox" is no paradox at all; it is due to both space and time being different for different observers (i.e. locations) within the theory of relativity. In the case of a black hole, this difference is drawn to its extreme. Time, as seen from outside, is standing still on the event horizon while it is ticking on as usual for an astronaut falling through it.

Even though weird things happen at the event horizon - nothing can come out if it ever goes in, and time itself and one of the three dimensions of space change places and meaning in the formulas - we now know that the Schwarzschild radius does not define a real singularity because the curvature is not infinite there. Instead, it is a "fictitious" singularity due to that the Schwarzschild coordinates originally used are not suitable right there. Therefore it is sometimes called a "coordinate singularity" as one may transform it away by simply switching to a new coordinate system. In the center, i.e. where $r = 0$, there is a "real" space-time singularity where the curvature really *is* infinite. This was originally taken as a sign that physics itself might be "squeezed out of existence" and ends there, but is nowadays usually considered as a sign that general relativity is no longer applicable at extremely high densities, that is: we have stretched Einstein's theory to, and beyond, its breaking point. It is believed that quantum mechanical effects become important before the singularity is created, and probably will prevent it from arising altogether. But so far it is just a guess because no detailed theory of quantized gravity, "quantum gravity", yet exists.

What happens inside the event horizon? The region, according to general relativity, is empty except for the singularity at the center; at least if one waits a while after the black hole formed and also neglects sporadic infalling objects, particles and radiation from the outside. Everything that crosses the event horizon will therefore inevitably be sucked into the singularity. A fair analogy is to view the space inside as a waterfall that constantly rushes toward the center faster than light. This means that a swimmer (light) can never get out because the current (space) moves inward faster than anyone can swim (the speed of light). And if light cannot escape, nothing else can either.

Since the original matter is crushed to nothing - a mathematical point at the center with zero volume - the matter really has disappeared from our universe. John Wheeler has called this "mass without matter" - the mass is left as the black hole still curves space-time outside (and inside) the event horizon, although the matter has actually disappeared.

An unexpected thing is that you can actually cross the event horizon of a (huge) black hole without knowing it, yet you can never leave and your future is certainly in the singularity at the center. How it feels when you cross the event horizon depends on the so-called "tidal effect", as it is the same effect that cause tides on earth. If you fall into a small black hole, the tidal effects are extreme, so that the part of the body about to enter is experiencing an effective force that can be billions of times greater than on the part of the body furthest out. This means that you are experiencing an enormous force of tension that tears apart the very atoms in your body. (Of course, death comes long before this.) The larger the black hole, the milder the tidal force at the event horizon and the giant black hole becomes unnoticeable - which does not prevent it from tearing you apart before you, only slightly later, fall into the singularity in its center.

In more realistic cases where it is not assumed that the original body was spherically symmetric one can show that the singularity will create chaotic oscillations in the curvature of space-time that grow to infinite strength as one approaches the singularity. This more physical solution (as compared to the idealized Oppenheimer-Snyder singularity) is called a BKL-singularity after its discoverers Belinskii, Khalatnikov and Lifshitz.

In 1963 Roy Kerr discovered a novel exact solution to Einstein's equations, which was later understood to describe a rotating black hole. The solution is more complicated, because it no longer has spherical symmetry, but it also has completely new and intriguing physical effects that do not exist for the Schwarzschild solution. The singularity is no longer a mathematical point, but a ring lying in the rotational plane along the "equator". Since the ring is a mathematical line of zero volume its density, and consequently the curvature, is still infinite – "singular".

In addition, the space-time outside is forced to rotate with the black hole, an effect known as "frame-dragging", roughly as if space-time consisted of viscous oil. Within a certain distance, the so-called "static limit", not even light itself can rotate in the direction opposite to the black hole, but is "dragged along" in the space-time rotation. The interesting point is that the static limit lies *outside* the event horizon. Roger Penrose discovered (1969) that this means that the black hole's rotational energy can be extracted by a mechanism now known as the "Penrose process". He calculated that there is a region around the black hole, the "ergo sphere", where you can put something in and get it back out with a higher speed, i.e. with a higher energy. That energy must come from somewhere and Penrose showed it comes from the rotation of the black hole itself.

It is believed that the Penrose process (together with galactic magnetic fields) for giant rotating black holes in galactic nuclei of very early galaxies explains the enormous relativistic "jets" - particle beams near the speed of light - emanating from quasars; the most remote, and therefore, brightest objects we know of today. Black holes, surprisingly, are incredibly efficient at converting matter to radiation, much more effective than, say, nuclear power, which only converts about 1% of the mass into energy. (The only thing more efficient than black holes is the annihilation of matter and antimatter, where 100% of the mass becomes radiation.) Nothing can come out from the region inside the event horizon, but before the material falls through it will heat up enormously due to friction and interactions between the incident particles, which means they emit copious amounts of radiation, such as X-rays. In 1971 the first candidate for a black hole was observed: The binary star system "Cygnus X1" where the dark (unseen) star/black hole has about 10 solar masses. This object was first identified by the space-based X-ray Observatory, "Uhuru".

The most general solution for a black hole was discovered in 1965 by Ted Newman (and collaborators) and is therefore known as the Kerr-Newman solution. A black hole can in theory have only three properties, mass (M), angular momentum (J) i.e. rotation, and electrical charge (Q). Knowledge of these three parameters gives a complete description. A black hole is thus the simplest object in all of classical physics. Even a speck of dust needs an almost infinite number of parameters to be described in microscopic detail, not to mention a giant star that create a black hole. But once it *is* created, it is extremely simple. John Wheeler has chosen to describe it as: "a black hole has no hair", by which he means that the black hole does not have any details beyond M, J, Q. If you know the mass, rotation and charge you know everything there is to know about the black hole, at least according to the purely classical theory of general relativity. What happens when quantum physics is taken into account (which has to be done because the most fundamental level in the universe today is described by quantum physics) you may speculate about yourself as no one else today knows...

Important people

Albert Einstein - Invented general relativity theory (1915), but was strongly opposed to the idea of black holes (singular solutions).

Karl Schwarzschild - Discovered (1916) a solution to Einstein's field equations describing an isolated, spherical, and unchanging mass that does not rotate.

Robert Oppenheimer & Hartland Snyder (1939) - Showed that the Schwarzschild solution, if all the mass M is inside the Schwarzschild radius, collapses into a black hole.

David Finkelstein (1958) - Discovered that the event horizon of a black hole acts like a one-way membrane; easy to get in, impossible to get out.

Roy Kerr - Discovered (1963) a generalized Schwarzschild solution in which the mass also rotates.

Roger Penrose - Showed (1964) by powerful global (topological) methods, which he himself developed, that the singularity cannot be avoided, regardless of how the collapsing matter is distributed and behaves.

Ted Newman - Published (1965), along with co-workers Couch,

Chinnapared, Exton, Prakash and Torrence, the most general solution possible for a black hole (at least in general relativity). The link to black holes was recognized only later, however.

Stephen Hawking - Discovered (1974), by approximately "gluing on" quantum effects on classical black holes (from general relativity), that they are no longer black, but radiate energy. The radiation, appropriately, has a "black body spectrum"; the radiated energy in a given wavelength band depends only on the temperature.

John Wheeler - named the black hole (1967) and coined the phrase "a black hole has no hair", meaning that (classical) black holes do not have any details, instead they are the simplest objects in all of classical physics.

Jacob Bekenstein - Discovered/guessed in 1972 that black holes must have entropy (disorder) proportional to the area of their event horizons. This was apparently at odds with the no hair-theorem.

Stephen Hawking - Hawking Radiation: When quantum physics is taken into account it seems black holes radiate energy at a certain temperature - the higher the mass of the black hole, the lower the temperature. This result was first derived in an attempt to refute the Bekenstein conjecture. Instead, it showed that Bekenstein was right. Entropy and temperature are intimately connected thermodynamically, and one can actually translate the laws of thermodynamics into similar laws for black holes.

Basic equations

The Schwarzschild metric (in spherical polar coordinates):

$$ds^2 = (1 - \frac{2GM}{c^2 r})c^2 dt^2 - \frac{1}{1 - \frac{2GM}{c^2 r}} dr^2 - r^2(d\theta^2 + \sin^2\theta d\phi^2)$$

The Schwarzschild radius (giving "fictitious singularity" in the metric above):

$$r_s = \frac{2GM}{c^2}$$

Gravitational red-shift:

$$z = \frac{1}{\sqrt{1 - \frac{r_s}{r}}} - 1$$

Static limit (written in "relativistic units" $c = G = 1$):

$$r_{stat} = M + \sqrt{M^2 - \frac{J^2 \cos^2\theta}{M^2}}$$

Hawking Radiation temperature:

$$T = \frac{\hbar c^3}{8\pi G k} \frac{1}{M}$$

7 CHAOS THEORY IN 15 MINUTES

The world is no Clockwork

Chaos is everywhere. Whatever systems we look at, most of them are intrinsically chaotic and therefore impossible to predict in detail. That we regardless have come so far in the exact sciences depends on the fact that we for centuries have ignored this and calculated in approximate ways, ways that in many cases have worked brilliantly, but sometimes not at all.

Modern researchers sometimes claim to be able to explain in detail phenomena as exotic as exploding giant stars, the first moments of the creation of the universe and its final fate. The reality is more sobering; one cannot even understand something as common as a dripping faucet, in detail, despite it being there right in front of our eyes.

It has recently become known that under certain circumstances, for almost any physical system whatsoever, the unavoidable errors in measured data, alternatively the external perturbations always present, will grow at each step forwards in time – and the error grows exponentially. After a short time nothing can any longer be said about the system's detailed behaviour. What constitutes a "short time" depends on what we are studying: a small fraction of a second for turbulent fluid flow, millions of years for planetary motion.

The phenomenon described above is usually termed deterministic chaos, or just chaos. This type of chaos can be summarized in two short lines:

1) There is an exact (deterministic) rule for predicting the future behavior, but...
2) In practice this becomes impossible as we never have the infinitely precise in-data that is required.

This means that even if we would know exactly how to calculate something, for example weather forecasts, it would do us no good as the observed values we use as our starting point at a specific time never can be precise enough for the calculation to follow the future weather. It would take measurements with infinite accuracy at every point in the atmosphere, obviously impossible.

The weather is thus a chaotic system, which means that we never will have detailed forecasts beyond two weeks into the future, regardless of how powerful computers will be constructed.

Brainwashed physicists

Chaos theory is today regarded as "self-evident" by many scientists. How can it be that it took so long to realize that so many systems can be chaotic? The reason for chaotic behaviour can be stated "nonlinear systems" (to be exact it is a necessary but not sufficient condition). Newton himself well knew that his equations, that describe general motion, could be nonlinear. The problem was that neither Newton, nor anyone else, could solve them except in rare special cases unless one made simplifying assumptions. The equations were "linearized", that is; the annoying nonlinear terms that prevented solutions to be found were removed. After roughly a century of using this procedure, the physicists had brainwashed themselves into thinking that nature itself was linear (even though it had started out as a simplifying mathematical trick). Everything was therefore, already at the outset, embedded in the classical laws of physics, both in Newton's equations and in Maxwell's equations for electromagnetism. The physicists had simply tricked themselves by only considering what could be solved by paper and pen, that is, only the solutions that could be written by means of a formula. By so doing, they also wiped away all possibilities to discover what today is called deterministic chaos; a system that looks like it is behaving randomly even though it is governed by (often) very simple equations. A solution in terms of a formula is simply too regular to be able to show chaotic behaviour.

Therefore, especially through the influence of prominent people like Newton's disciple Laplace, nature was seen as a perfectly predictable clockwork that in the future one would undoubtedly come to know in complete detail. In some cases, under "nice" conditions, this linearization works wonderfully, but in some cases it does not…

Lorenz sees chaos

The mathematician and physicist Poincaré was the one who saw the first glimpses of chaos already at the end of the 19th century, but he was mostly horrified by what he saw. It had to await the advent of modern computers in the 1960s before one realized that nonlinear and chaotic behaviour was more the rule than the exception. The one, who first discovered this, through computer simulations, was the American theoretical meteorologist Edward Lorenz. The story of Lorenz' chaos is a classic in the development of science, as almost all great breakthroughs are discovered by chance. It started with his construction of a very simplified model, a numerical computer simulation of the development of a weather system. He punched in his input data, crunched the numbers, and got a result. When he wanted to rerun the computation he started from an intermediate point in the first calculation. Inadvertently, he discarded the last few decimals, how could they make any difference anyway? According to the prevailing thoughts in the physics of that era, the resulting curve should deviate only slightly from the previous one – but instead a completely different curve was generated, without any resemblance to the first. Lorenz had seen chaos. It took computer simulations before the mental brainwash of scientists could be lifted.

Theory of a dripping faucet

But the dripping faucet we already mentioned surely cannot have an extraordinary explanation, or can it? Everybody surely knows how a dripping faucet behaves? That was exactly how people reasoned before Robert Shaw wrote his whole doctoral dissertation on the behaviour of these drops. You can, yourself, try out the chaos generated by a dripping faucet. If you turn on the flow just a little bit, so you get a nice steady flow, you have created what is known as laminar flow which is relatively simple to describe mathematically. Now open it fully, the laminar flow turns to turbulent flow which is virtually impossible to describe in detail. Now, instead do the opposite, reduce the flow until not even one drop is created per second. The dripping will now be periodic; every drop arrives after the same time interval, drip-drip-drip… Slowly increase the flow and you can create a two-periodic dripping; drip-drip-drop-drip-drip-drop… Depending on your faucet and your own skill you might be able to create a few more multi-periodic drop-sequences, that is, ones that take increasingly many drops before they repeat. But above a certain limit for the flow no exact duplication of the sequence will ever take place regardless of how long one waits – the dripping has become chaotic.

Robert Shaw used a more exact "faucet" (a needle valve) in his studies and a laser beam to measure the time interval between succeeding drops, but in principle he did the same thing that you (may) just have done. In fact, Shaw's investigation was part of the revolution during the 1970s that now goes by the name chaos theory.

The reason why the drops behave so weird is due to the nonlinearity of the system. If and when the drop is released is determined by a complex interplay between the force of gravity and the electromagnetic forces (surface tension, etc.) that retain the drop and keep it from falling. Furthermore, a drop will send a shock wave through the remaining water and influence when the next drop will fall. For slow rates the shock dissipates and one gets the highly predictable periodic behaviour, only dependent on the flow of water. For higher flow rates the shock wave can enhance or suppress the "natural" flow. Eventually, the interaction becomes so strong and complicated that nothing can be said about the detailed time of arrival of the next drop.

In the chaotic region (high flow) it does not matter how good and extensive information we have about the historic drop-behaviour. One may have collected data for 100 000 (or even infinitely many) earlier drops without it helping one bit in determining when the next drop will fall.

In fact *all* physical systems are nonlinear. But only in some cases will the system be chaotic. The Lorenz model, for instance, can be made to behave completely regularly (non-chaotic) for some choices for the parameters σ, ϱ and β. The heart and the brain are two other examples of chaotic systems. It is an evolutionary advantage. They need to be able to change states very quickly, for example if you meet a bear in the woods, and a chaotic system can change states much faster than a more stable, linear system.

Nowadays one also introduces nonlinearities on purpose. For instance modern fighter planes are constructed to be intrinsically unstable, and would fall like rocks if the stabilizing computers would be taken out. One of the reasons is that a chaotic object is easier to influence, and the fighters can turn steeper because the design in itself is directionally unstable.

A system can thus behave simply and non-chaotic under some circumstances and chaotic under others. In our example with the dripping faucet we could regulate this by just altering the flow of water. Chaos theory has led to the conclusion that:

1. Seemingly complicated and "random" behaviour can be the result of very simple, but nonlinear, laws.
2. Nonlinear systems are the rule; linear systems are approximate descriptions that only work some of the time.
3. In the chaotic region the possibility for detailed prediction disappears very quickly.
4. There is plenty remaining to be discovered, both hidden in "known" laws of physics, and in the natural laws we have not yet found.

There are for instance astrophysicists that claim to know almost "everything" about a supernova explosion, the final phase of the life of a giant star. But as a supernova has thousands of input parameters, and we have seen that even the dripping faucet with only one parameter (the flow) is not fully understood, we realize that we should be sceptical to all such announcements. Science is never a finished chapter. Today's "truth" is only a stepping stone towards the, probably very different, world view of the future.

Linear and nonlinear

The concepts linear and nonlinear have, somewhat simplified, their origin in how input-data (measurements) and output-data (predictions) are related. For linear systems an error in input by one per cent means that the error in output will be roughly of the same magnitude. For a nonlinear system, on the other hand, the error can become much larger and grow much faster. When the error becomes bigger than the quantity we actually are interested in, all power for detailed prediction of the future behaviour has been lost.

Two types of chaos

Usually, one distinguishes between two types of chaos: with friction (dissipative) and without (non-dissipative). "Dissipative" means that energy disappears from the system while the energy in non-dissipative systems is conserved. The famous "strange attractors" can only arise in dissipative systems, as the only way for the system behaviour to approach an attractor is if energy is being "bled off". Although attractors do not exist for non-dissipative systems, the trajectories (in phase space) can become infinitely entangled even though the total volume cannot change. This is called Hamiltonian chaos. In both cases, that is, with and without friction, the governing equations must be nonlinear for chaos to be possible.

Strange attractors

Strange attractors live in "phase space", which really sounds more complicated than it is. For a plane pendulum, phase space is two-dimensional and the two coordinate axes are given by the pendulum's angle and its velocity. For a pendulum with friction the final state, i.e. the state the system will become attracted to in phase space, is the origin where both the amplitude and the velocity are zero. If we connect an outside driving force to the pendulum, the final state can instead be that it continues to oscillate at the same rate, with the same amplitude, indefinitely. The attractor has then become a closed trajectory, a "loop" in phase space.

If the description of the system requires at least three dimensions, or coordinate axes, and the equations are nonlinear, then exotic, chaotic, "strange attractors" can arise. These have "fractal dimension", in contrast to the integer-dimensional attractors of pre-chaotic systems, for example the pendulum with friction (point attractor with dimension zero), and the pendulum with driving force (line attractor with dimension one).

Important People

Henri Poincaré - Poincaré is regarded as one of the greatest mathematicians in history. He saw the first signs of chaotic behaviour, in the end of the 19th century, "a nightmare". This happened in connection with Poincaré winning a competition (1889) announced by the Swedish king Oscar II, with purpose to decide if the solar system was stable or not.

Edward Lorenz - Lorenz found (1960), by coincidence, the first concrete example of chaotic behaviour in a very simplified model for the weather. No simple periodic motion occurs, instead the system settles unto a "strange attractor" (1963) which implies that the behaviour never exactly repeats. Before the finding of Lorenz all attractors had been simple geometrical objects. Points – where the final state is completely still, for example as for a pendulum with friction that eventually comes to a halt. Closed trajectories – corresponding to periodic motion of the system, as for a pendulum with a driving force, e.g. as in a pendulum clock. Surfaces – representing multi-periodic motion. (Strictly, surfaces are generated only when the quotient of the different oscillation periods are irrational numbers.)

David Ruelle - Ruelle and his collaborator Floris Takens for the first time (1971) connected turbulence in fluid flow with the concept of chaos.

Robert May – (1976) May reviewed chaos in the "logistic map" (a very simplified model for the number of individuals in a population of animals).

Mitchell Feigenbaum - Feigenbaum found (1978) that many different chaotic systems show similar behaviour. In other words there exists in some cases "universality" in chaos, not obvious to a casual observer.

Benoit Mandelbrot - Mandelbrot discovered "fractals" (1975), infinitely jagged geometric figures, in part by his mathematical studies of coastlines. He realized that the length of the coast is not decided once and for all, but depends on the length of the ruler used. Later, fractals became intimately linked to chaos as chaotic "strange attractors" are fractals in phase space.

Robert Shaw - Shaw (1984) showed that even very simple systems, like a dripping faucet, can exhibit very complicated, chaotic behaviour.

Ilya Prigogine - Prigogine showed that systems forced far from equilibrium can exhibit self-organizing characteristics. For this he got the Nobel Prize in chemistry in 1977. Prigogine also made great contributions to the subject of deterministic chaos, and its connection to quantum physics.

Basic equations

"Lorenz' equations" constitute an extremely simplified convection-model for the weather.

For some values of the constants σ, ρ, β the system is chaotic:

$$\frac{dx}{dt} = \sigma(y - x)$$

$$\frac{dy}{dt} = x(\rho - z) - y$$

$$\frac{dz}{dt} = xy - \beta z$$

Lorenz' "strange attractor" which the system (in the chaotic region) given by Lorenz' equations approaches, and then moves along, independently of where it initially starts. The dimension of the attractor is "larger than a line" but "less than a surface" which is impossible in classic geometry, i.e. before fractals:

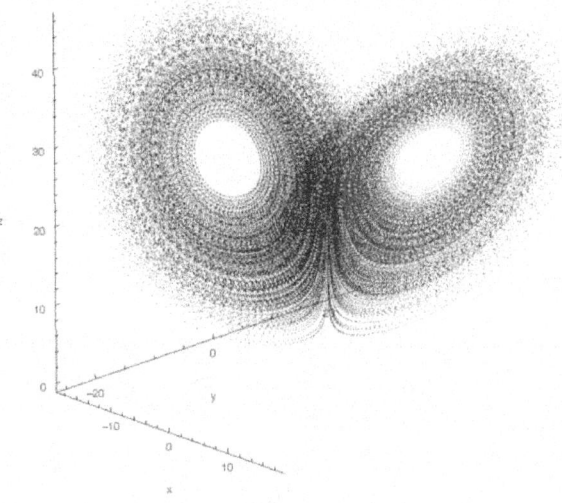

The very simple generating function for the Mandelbrot set in the complex plane. The Mandelbrot set (black regions in the figure) is defined by starting points *not* flying away to infinity as the generating function is applied over and over (iterated). The "coastline" of the Mandelbrot set is a fractal; regardless of how many times it is magnified it is always as jagged:

$$f(z) = z^2 + c$$

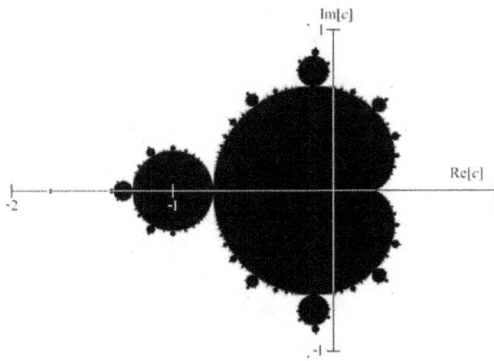

The logistic equation" is another iterative model where the next number is given by the preceding one. In a population model the numbers x stand for the number of individuals in successive generations. Despite its deceptively simple appearance it hides a very complex behaviour:

$$x_{n+1} = rx_n(1 - x_n)$$

A "bifurcation diagram" for the logistic equation is shown. For small values of the constant r the logistic equation exhibits a point attractor (that, for example, always drives the solution towards $x = 0$ as $r < 1$), for higher values of r, x first jumps (oscillates) between two values, then between four, eight, sixteen, ... When $r > 3.57$ the behaviour becomes chaotic and jumps randomly between different values, filling all possibilities as $r = 4$:

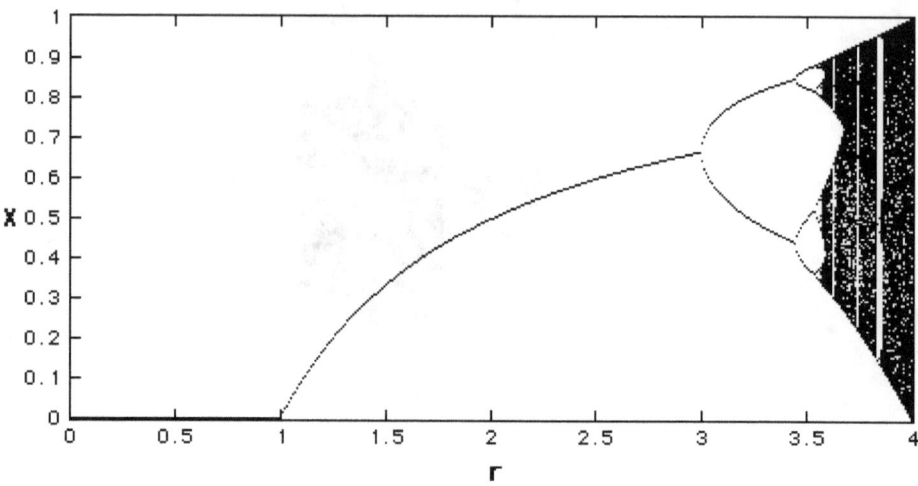

The "Feigenbaum constant", δ, shows that there can be "universality" in chaos. The logistic equation and similar one-dimensional equations are, despite the chaotic behaviour, predictable in the sense that they behave in the same fashion. There is an underlying order in the chaos:

$$\delta \equiv \lim_{k \to \infty} \frac{a_k - a_{k-1}}{a_{k+1} - a_k} = 4.669201609...$$

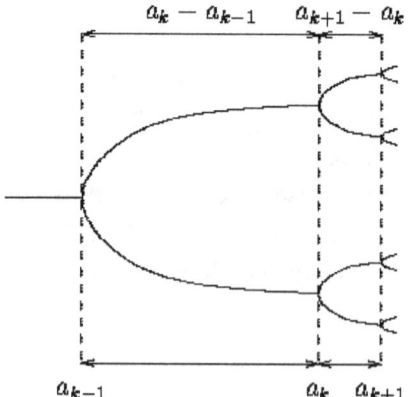

8 STRING THEORY IN 15 MINUTES

The theory of Everything… or Nothing?

Introduction

Physics is the science of nature. A theory that has not yet been tested against the phenomena in nature itself is not (yet) a physical theory. There are countless examples of historical conjectures about how nature "should" behave having been proved wrong once they could be tested experimentally. There is no reason to believe that it will be different today, or in the future. Always keep this in mind when you hear about the latest "revolutionary" - but untested - theories of the universe's mysteries.

String theory has as one of its main objectives to solve the biggest challenge in theoretical physics; to reconcile quantum mechanics with general relativity, that is, to construct the lacking theory of quantum gravity. The second goal is to unite all of the four known forces of nature into one; to show that all the force-carrying particles (bosons) in fact can be generated by different vibrations of the fundamental string. The third objective is to also describe all matter particles (fermions) as vibrations of the string. If these three goals were met we would have a "theory of everything" because it would explain the space-time itself (that is, gravity) and all particles and forces in it. The fundamental building blocks of string theory are therefore one-dimensional vibrating strings, but deeper studies of string theory has proven to describe not only strings but also point-like objects and so-called (mem)branes with higher dimensionality.

The problem is that all this applies only for the most extreme conditions imaginable, for example when the universe just had been created. In today's universe, the various forces have long since split apart to four separate forces, and the particles have condensed out to all the distinct varieties we see in particle physics experiments. To re-create the united original state using today's technology would require a particle accelerator larger than our entire galaxy. String theory therefore, at present, resemble the philosophy during the Dark Ages in Europe debating how many angels can fit on the head of a pin, while astronomers in China simultaneously made detailed observations of real giant exploding stars, supernovae.

It is also not known whether string theory could describe a universe with exactly the forces and particles that actually exist, that is, if one can deduce from the theory the quantum fields which represent the experimentally tested Standard Model of particle physics. It is also unclear how much freedom to choose those details the theory allows - if string theory is to function as an ultimate "theory of everything" the standard model should pop out as a *unique* solution.

Another problem is that string theory so far is formulated only approximately. So there is no unique string theory, but rather countless different ones.

More recently, string theory also has been subjected to other criticisms, mainly by Lee Smolin, who himself is an old string theorists. The main argument is that the theory is more a product of sociological mechanisms than of scientific research. Smolin argues that many of today's leading researchers indirectly have chosen to work with string theory because today one "must" choose it if one wants to pursue an academic career in theoretical physics, where almost all jobs in the U.S. and abroad is about string theory.

String theory's origin

String theory has been around since the late 1960s, originally designed as a theory of how protons, neutrons and other nuclear particles interact. With the development of QCD - Quantum ChromoDynamics - the quantum field theory that describes the "color" force between quarks, this ambition died, and all but a handful of researchers abandoned string theory. The renewed interest in string theory came in the 1970s as it was realized that a previous undesirable feature could instead become an asset. It had long been known that string theory automatically contains many additional particles, including a massless spin-2-particle. There was no room for this in the force between protons and neutrons, but if reinterpreted as a Graviton, i.e. the particle of the gravitational force, string theory was suddenly a potential candidate for a unified "theory of everything". This is the great vision that drives string theorists, the possibility to find a single theory that everything in the universe is described by. Not only the particles and forces, but even space-time itself may be described by this theory, or rather its modern successor, M-theory. The idea is that there is only one type of string, nature's most fundamental element. Depending on its undulations, the result is an electron, a quark, a photon, or any matter or force particle. Because there are infinitely many ways in which the string can oscillate there are also, according to string theory, an infinite number of particles. These should all have at least around a "Planck mass", about the same mass as a bacterium, which is very tiny for us, but enormous for an elementary particle. This is one of many unsolved problems in string theory; why the particles we actually *see* in the experiments have the masses they have, and are almost massless compared to the natural string theory unit of mass, the Planck mass. String theory is now fully integrated with supersymmetry (SUSY), and therefore it is called superstring theory.

One should remember that string theory so far is just an exciting possibility; there is *no* evidence at all that string theory is something that describes nature. No physical phenomena have been explained so far by string theory, which in itself is a bit odd given that it has existed for some 40 years and that thousands of physicists in the world are working on it. In fact, string theory today is not a physical theory at all, because it has not been tested against experiments, and probably will not be able to be tested by direct experiments in the foreseeable future. However, there are proposals for indirect tests. One of the most promising is to study the subtle signs of strings in cosmological observations. Although the effects of the strings are ridiculously small they are cumulative with distance. If we are observing objects and phenomena at very large distances (comparable to our entire visible universe), there is some chance to see these effects already with today's technology. Nothing has yet been seen...

In superstring theory, the big "showpiece" otherwise mentioned is the successful calculation of the entropy, or "disorder", for a black hole, i.e. the black hole's temperature, thereby reproducing Hawking's half-classic result from 1974 in which he "glued" quantum effects onto a classical black hole solution of general relativity. But the temperature of a black hole has never been measured. Therefore, this "success" is just an example of how one compares the outcome of one hypothetical theory with results from an even more hypothetical theory, no real physics.

String theorists themselves have also long recognized that superstring theory cannot be the final answer. Among other things, string theory cannot explain the origin of space and time, because the strings are moving in a "background" of classical space-time that is borrowed from Einstein. It is therefore impossible from the outset to realize the dream of explaining how gravity (curved space-time) arises from string theory because it is only known how to deal with strings if they are small "disturbances" on a fixed space-time background. In addition, there are many different string theories and all of them seem to be equally good, no great asset for a theory where the idea originally was that it would be unique and explain everything. Therefore, they are now looking for the theory "behind it all", which is called M-theory. Unfortunately, not much more than the name exists as of yet, and there is not even agreement on what the abbreviation M stands for.

Extra dimensions

The original "bosonic" string theory called for 26 dimensions to make sense mathematically, whereas the more modern supersymmetric version needs 10 dimensions, while M-theory has 11 dimensions. This seems strange to say the least, given that the universe has only four dimensions (three spatial dimensions and one time dimension). Theodore Kaluza and Oskar Klein, as early as the 1920s, suggested additional dimensions scrunched up so small that they are invisible to us, in an attempt to reconcile electromagnetism with general relativity - that is, all the known forces in those days - into a single "Grand unified theory". So the string's extra dimensions were then also assumed to be "compactified" to the "Planck length" (10^{-35} meters, 10^{20} times smaller than an atomic nucleus). However, there was, and still is, no explanation of how and why our usual four dimensions, were "left over" without compactification...

In recent years people have also begun to examine the possibility that there could be much larger extra dimensions, but that gravity alone has access to them. It would explain why gravitation, as seen from our four-dimensional point of view, is so weak. Actually gravity can then be comparable in strength with the other forces; it is just diluted, sent into all the other extra dimensions. Large extra dimensions are interesting for many reasons, including dark matter, but suffers the same deficiency as string theory itself, the idea has not yet been tested experimentally. By indirect tests using Newton's law of gravitation it has been concluded that the "large" extra dimensions (if they exist) must at least be smaller than 0.1 millimeters, the limit for today's best experiments.

Physics? ...

Even professional physicists sometimes mix up our *description* of nature with nature itself. Physical laws are our attempt at a simple, compact description and summary of how nature has been shown to behave, at least as far as we are now able to test. A pretty good analogy is that laws of nature are like maps. Nobody would think of the idea that the map itself is nature, but that is sometimes what happens in physics. The map is useful because it can easily and graphically summarize the most important aspects of nature. A map that is as complete as nature itself would not make any sense at all, as we then just as well could study the original, nature, directly. Therefore, one can question the modern "Holy Grail" of physics, a Grand unified "theory of everything". A map will be updated whenever new facts become known. The same will probably forever be true for the laws of nature; they must be rewritten as more facts gradually become known. Of course one should still continue research in superstring theory, but today it has a disproportionate share of all theoretical physicists, especially as it is a theory that may not have any connection to nature and the real universe at all. Remember, physics is the study of nature, *not* the study of theories. As so aptly stated: "The history of physics is littered with the Corpses of Dead theories."

About...

SUSY is an abbreviation of "supersymmetry". This is a symmetry mathematically discovered in the early 1970s which in an ingenious way generalizes spacetime-symmetry in Einstein's special relativity. Supersymmetry is embedded in superstring theory. SUSY says that every particle must have a supersymmetric partner. The main reason for introducing SUSY is that it cures certain mathematical problems (including infinities) in theories without SUSY. However, no supersymmetric particle has yet been found.

Important people

Gabriele Veneziano - worked in the late 1960s on a theory of the strong nuclear force - the "dual resonance model" - to explain the large amounts of incomprehensible experimental data streaming out of particle accelerators. As quantum chromodynamics (QCD) was developed in the early 1970s, and was shown to be much better at describing the strong nuclear force, interest in the dual resonance model waned. Veneziano's model instead came to be the starting point for today's string theory (1968).

Yoichiro Nambu - realized that Veneziano's theory describes quantum strings (1970).

Pierre Ramond – introduced for the first time supersymmetry in string theory. Superstring theory is born (1971). Supersymmetry can be described as matter (fermions) and forces (bosons) being related to each other; it says among other things that matter particles and force particles should have the same mass. Since this relationship is not seen in experiments supersymmetry must now be broken. No one to this day knows exactly how.

John Schwarz & Joel Scherk - realized in 1974 that an unwanted massless spin-two string-excitation, which caused problems for string theory as a strong nuclear force (because the nuclear force would then have infinite range, contrary to observations), rather "automatically" may include gravity - the excitations identified with the Graviton, gravity's hypothetical force-carrying particle. This was the start of the dream of a unified "theory of everything", an amalgamation of all four forces of nature into a single "primordial force".

Michael Green & John Schwarz - 1984 showed how some serious mathematical problems (so-called "anomalies" that would make strings useless for calculations) in string theory could be solved, but only if the dimensions were expanded from the known four to ten. This marked the start of the explosion of interest in string theory, before that only a handful of enthusiasts worked on it.

Edward Witten - guessed in the mid-1990s that the many variants of string theory can be reconciled in a "Super-duper"-theory (with one more dimension, i.e. a total of eleven), called M-theory. There is not much more than the name as yet, not even agreement on what the letter M stands for.

Joseph Polchinski - discovered in the mid-1990s that string theory actually requires/predicts membrane-like objects, called p-branes ("pea-brains") and D-branes, with dimensions greater than a string (which has dimension one, because it has only length).

Juan Maldacena - suggested in 1997 that string theory and "normal" quantum field theory of a special type can be two aspects/descriptions of the same thing. This prediction has been called "AdS/CFT correspondence".

Basic equations

A free "Bosonic" string, with string tension α, which moves through flat spacetime sweeps out a "world surface". Points on the two-dimensional surface are described by the two coordinates τ and σ. The "action" S (which quantization of the string is based on) is given by

$$S = -\frac{1}{4\pi\alpha'} \int_{-\infty}^{\infty} d\tau \int_0^l d\sigma (-\gamma)^{1/2} \gamma^{ab} \partial_a X_\mu \partial_b X^\mu$$

For a super(symmetric)string the space-time coordinates X, the "bosonic" quantities in the theory, are related to the fermionic fields ψ via supersymmetry

$$S = -\frac{1}{4\pi} \int_{-\infty}^{\infty} d\tau \int_0^l d\sigma (-\gamma)^{1/2} (\frac{1}{\alpha'} \gamma^{ab} \partial_a X_\mu \partial_b X^\mu - i \overline{\psi}_\mu \rho^a \nabla_a \psi^\mu)$$

9 QUANTUM GRAVITY IN 15 MINUTES

The greatest unsolved problem in theoretical physics is how to make gravity and quantum mechanics coexist under the same theoretical "umbrella". In other words, how to integrate general relativity and quantum physics in a single theory. Many people believe that quantum physics should be maintained more or less unchanged, and in this merger general relativity should be modified. But some, with the renowned British mathematician and physicist Roger Penrose as main proponent, instead believe that the general theory of relativity is the more fundamental one and that quantum physics has to be reformulated to make the unification possible.

Quantum gravity is required to make the whole of physics logically coherent, and is hoped to answer many unsolved fundamental mysteries: How did the universe begin? Did time exist before its origin? What exactly happens in a black hole (the singularity is almost certainly a sign of the classical theory breaking down and becoming unusable, not that nature really is singular). Does information disappear in a black hole? Do black holes even exist? What is dark energy? Does *it* exist? How and why do the over twenty free parameters in the standard model of particle physics arise (which today have no explanation at all)?

The problem is that quantum physics and general relativity already overlap each other's domains, but do not fit together. The other three of the four known forces of nature (weak and strong nuclear, and electromagnetism) are all described with full consideration taken of quantum physics, and are included in the so-called standard model of particle physics. Although the coexistence of special relativity and quantum physics is not completely peaceful, one has found a way to deal with them simultaneously in the subject called "quantum field theory". Unfortunately, the same quantum-field-theory-methods have never worked to combine *general* relativity with quantum physics. The main problem with quantum gravity, from a purely scientific point of view, is that one cannot do the experiments required. Since quantum gravity effects are impossible to observe today because the energy needed is way above what will be available in the foreseeable future, controlled active experimentation and comparison with the theoretical predictions (in the few cases where one even can calculate them) is impossible. For example, a particle accelerator based on current technology needs to be larger than our entire galaxy in order to test these tiny effects. This means that quantum gravity – today - is not really a science in the traditional sense. No experimental input exists which can be used to check theoretical concepts, and historically we know that theoretical "progress" then usually is in the wrong direction. There are suggestions of how to indirectly observe the quantum gravitational effects ("large" extra dimensions/deviations from Newton's law of gravitation, light from gamma-ray bursts, variations in the speed of light, mini-black holes in particle accelerators, the effects of gravitational waves, etc.), for example, in cosmological observations, but so far no one really knows how.

Today there are two main competing ideas which are hoped to lead to a working theory of quantum gravity: Superstring/M-theory and Loop Quantum Gravity:

Superstring theory/M-theory

String/M-theory has its origin in particle physics; it is the particle physicist's way of attacking this holy grail of physics. As early as 1930 Leon Rosenfeld for the first time wrote down the equations that apply when gravity is analyzed as an "ordinary" quantum field. This was, in fact, just after Paul Dirac had created quantum field theory, which was an extension of his own relativistic quantum equation - the Dirac equation - for the atom. The problem was that at that time, quantum field theory could not yet be used for detailed calculations, as it always predicted infinities, of course in total contradiction to known experimental facts. Quantum gravity was no exception. During the 1940s, after a lot of hard work, one succeeded in developing a functioning (but far from perfect) solution, the so-called "renormalizing" of quantum electrodynamics (QED). QED thus became the first working quantum field theory, in perfect accordance with all experimental results. Richard Feynman, Sin-Itiro Tomonaga and Julian Schwinger were eventually (1965) awarded the Nobel Prize for their work in making special relativity and quantum physics "peacefully coexist" in the electromagnetic interactions.

In the 1970s, the last nail in the coffin for the hope that quantum field theory might be used for quantum gravity was hammered in; when it was shown that the resulting theory is non-renormalizable. In other words, one would need an infinite number of renormalizing-constants (compared to only two in QED) to calculate results in quantum gravity, which of course meant that the theory in practical terms was meaningless.

The origin of the infinities that arises is that the particles are points and thus can interact with themselves at exactly the same location. The closed "loops" of force particles that arise as a consequence cause the infinities. In the 1970s it was briefly believed that the introduction of supersymmetry could remove the infinities, and people began to study "super-gravity", where each matter-particle got a super-symmetric force-particle partner, and vice versa, with the hope that the infinities would somehow cancel each other out. That hope was soon dashed, however.

How can string theory help? Because the fundamental entities are assumed to be strings, not points, the infinities automatically disappear, at least for each term in the series of successive approximations. But when one adds up all contributions of the entire series (infinitely many) the end result is still infinite, an annoying unsolved problem in string theory. In string theory all particles and forces are also assumed to be just different vibrations of the more fundamental string - the real constituent of everything in the universe. However, string theory has yet only approximate solutions, assuming a fixed spacetime background where strings represent small, insignificant "perturbations" in the spacetime geometry. This disqualifies it, from the start, from being the theory to correctly describe quantum gravity, since that would require that the strings themselves "weave together" spacetime in a string tapestry, not merely as corrections to a classical (non-quantum mechanical) solution. So even string theorists no longer believe that strings is the final answer. Instead, it is thought that an even more fundamental theory, called M-theory, underlies it all. Unfortunately there *is* no M-theory yet, but one guesses it is a theory of membranes, vibrations in "drum skins", of two dimensions or more. M-theory is sometimes jokingly referred to as "the theory formerly known as Strings"...

Another attractive feature of string theory is that it has the potential to unite both particles *and* forces into a single package, but so far it is only a theoretical possibility because no one has shown that the (experimentally) known particles and the three non-gravitational forces result from string theory.

If M-theory can handle both this *and* is able to provide a working theory of quantum gravity one has achieved what string theorists promised already in the 1980s, a Grand unified "primordial-theory" of everything, perhaps even summarized in a single formula. However, this is still only a glint in Ed Witten's eye, the string/M-theory über-prophet...

Loop quantum gravity

Loop quantum gravity (LQG) has its roots in the general theory of relativity, not in particle physics. Using the so-called "Hamiltonian" formulation of relativity Dirac, and later Arnowitt/Deser/Misner and DeWitt, showed how to quantize gravity without taking the path over quantum field theory. Later, in the 1980s, Ashstekar discovered new variables by which one could reshape the Wheeler-DeWitt equation, the fundamental equation in this theory, so that it was solvable. Smolin and Rovelli, in the 1990s, interpreted these solutions in physical terms, based on loop-quantization (using a method that Ken Wilson formerly had developed for particle physics/condensed matter physics) - roughly one can say that space-time, according to LQG, at its fundamental level consists of tiny loops that weave together the classical geometry as described by Einstein's general relativity.

Unlike string theory, the formulation of LQG is not approximate. Nor is there a need for extra dimensions (in contrast to string theory's ten and M-theory's eleven), the usual four (three space + time) are more than enough. Thus, LQG does not necessarily give a "theory of everything"; instead it focuses on solving the riddle of quantum gravity alone. Although LQG is significantly younger as a research field than string theory, it has reached more analytical results, at least mathematically. 1. It is an exact (i.e. non-approximate) quantization of space, where both area and volume are quantized, i.e. have "minimal, irreducible elements" at the Planck level, in contrast to classical geometry, where everything can be divided into infinitely small parts. 2. One can derive the entropy ("randomness", basically the temperature) of a black hole from fundamental principles and the result is consistent with Hawking's semi-classical formula from 1974 (also string theory claims this as one of its few fundamental results). 3. The singularity of the big bang is replaced by a non-singular "bounce".

However, there still remains much work to show that the theory is logically consistent, i.e. does not contain internal contradictions and/or obvious physical pathologies. Neither has it yet been possible to show that general relativity results from LQG-theory as a low-energy approximation, which is a requirement for LQG to have a chance at being physically correct.

Moreover, nobody knows how to test its predictions against phenomena in the real world, something it has in common with all other known candidates for quantum gravity.

Other ideas

Twistors were invented by Roger Penrose in the 1960s. At the beginning they were based on the notion of quantum mechanical spin, but were later expanded in other directions. This idea is not yet as well-developed as the two main approaches above (although both currently use aspects of twistors) and is therefore referred to as a model, rather than a theory.

Non-commutative geometry is a generalization of ordinary geometry, attempting to incorporate a fundamental property of quantum mechanics; that the *order* of certain operations are of great significance (the operations do not "commute"). In particular, the French mathematician Alain Connes has developed this approach. Today, both string theory and loop quantum gravity uses some mathematical techniques from non-commutative geometry.

There are also a host of other ideas by, for example; David Finkelstein, Charles Misner, Tullio Regge, Rafael Sorkin, Gerard t'Hooft, Lee Smolin, Stephen Hawking, James Hartle, Renate Loll, Chris Isham, and others...

What we "know" today

What we "know" today will be proven wrong tomorrow. Not in the circumstances that have been tested and are well-known, but when one looks at more and more extreme phenomena, all current theories are known to break down. Newton's physics works for normal everyday occurrences, but is completely inappropriate for microscopic phenomena, and also for common objects with relative speeds approaching that of light. The need to understand the atoms meant that Newton's laws of motion were superseded by quantum mechanics. High relative velocities were resolved by special relativity. The combination of quantum physics & special relativity (quantum field theory) was needed to explain the creation and annihilation of particles and antiparticles, which were first seen in cosmic radiation in the 1930s, and today is commonplace in particle accelerators. The relation between quantum physics and general relativity is still an unsolved problem, but even when that mystery eventually is solved, sooner or later a theory that generalizes and corrects even that will probably be required. It seems that fundamental physics is an endless adventure. For every mystery we solve, nature has many new in store for us.

The only thing certain is that tomorrow's physics, and our understanding of the universe, will *not* be what we today expect it to be. Two historic examples may illustrate the point: the existence of atoms was proven just a hundred years ago, and at that time one thought that the Milky Way was the entire universe...

All science is really an interconnected *whole*. The main reason that man divides the universe into small compartmentalized subject areas is that we simply cannot (yet?) understand everything at once. We have, after all, just a primate's brain... In nature there are no artificial divisions into physics, chemistry, biology, geology, astronomy, cosmology, etc., they are all a result of man's limited comprehension.

Einstein's dream was to describe the whole of nature in one single theory; that dream is still not realized... The difficulties are:

- Experimental - hitherto impossible to get clues from experiments/observations.
- Mathematical – the mathematics may not exist or is too difficult.
- Creative – a "new Einstein" may be needed.

Important people

Leon Rosenfeld - The first person who tried to quantize gravity (1930) by using the newly invented quantum field theory.

Peter Bergmann - Einstein's disciple, who built the first real research group to seriously study quantum gravity.

John Wheeler - Developed quantum gravity, from being an obscure fringe business, into a respected research field in the same spirit as other physics (Wheeler had previously been closely involved in the development of nuclear physics in the 1930s). Introduced a physical approach that basically had been missing before (when quantum gravity mostly was seen as a mathematical exercise). Almost all of today's researchers in quantum gravity have Wheeler as their "godfather". He derived, among other things, the Wheeler-DeWitt equation, the main equation in the so-called Hamiltonian or "canonical" quantum gravity.

Richard Feynman - In the 1940s solved the problem of infinities in quantum field theory (by, in his own words, "sweeping the problems under the carpet"). Then, in the late 1950s thought - as gravity is the weakest force of all - he would quickly and easily be able to solve the riddle of quantum gravity in the same way. It turned out to be impossible, but led Feynman, in the process, during the 1960s to come up with several techniques that subsequently proved indispensable for the modern "gauge theory", which today's particle physics is based on.

Bryce DeWitt - Listened to a talk in which Feynman described how quantum gravity might be solved at first and second approximation, and during the years that followed generalized the methods to arbitrary approximation level. Since gravity proved to be "non-renormalizable" (which means that the methods used to "sweep the infinities under the carpet" do not work) this still did not give a general solution. Also derived the Wheeler-DeWitt equation.

Paul Dirac - Made early, important contributions to the theory of quantum gravity, especially in the Hamiltonian formulation.

Reser Arnowitt, Stanley Deser, Charles Misner - invented a powerful, elegant formalism with which to treat (Hamiltonian) quantum gravity in a manner analogous to normal "simple" quantum mechanics (replace generalized momentum with quantum operators).

Roger Penrose - Became very famous for his "singularity-theorems" in general relativity in the 1960s. These were made possible as Penrose developed completely new and more powerful mathematical techniques than previously used, which Stephen Hawking then used to prove that a (classical) Big Bang model of the Universe must have started from a singularity, an infinitely compressed initial state. Penrose has since 1960 developed his own independent path towards quantum gravity by using the so-called "twistors".

Stephen Hawking - Wheeler's graduate student Jacob Bekenstein conjectured that a black hole should have an entropy, a temperature, because black holes would otherwise violate the second law of thermodynamics. Hawking initially believed that a black hole could not have a temperature, as it ought to devour all the radiation. He managed, through "gluing" quantum effects on a known geometry of a black hole, to produce an approximate formula (in 1974), which was consistent with Bekenstein's "guessed" formula. The temperature of this "Hawking radiation" is inversely proportional to the mass of the black hole, which means that mini-black holes (if such can be formed) would explode in a shower of gamma rays. Since this was the first concrete example of an effect of quantum gravity it was seen as a major step towards a precise future theory of quantized gravity. Hawking radiation has never yet been detected observationally, i.e. in reality.

Lee Smolin and Carlo Rovelli - Developed the modern, most promising variant of the "canonical" (Hamiltonian) quantum gravity, called loop quantum gravity (1990). It has its origin in exact methods from mathematical physics and is more stringent (mathematically strict) than string theory. Many of the results are therefore mathematically entirely accurate (unlike the case with string theory, which is an approximate theory); the uncertainty instead arises in the interpretation of what the solutions actually mean physically.

Basic equations

Wheeler-DeWitt equation:

$$H\psi(\alpha,t) = 0$$

Loop quantum gravity:

$$H = \int N \cdot tr(F \wedge \{\overline{V}, A\})$$

10 BONUS: PREONS AND PREON STARS

Introduction

Compact objects are astronomical bodies compressed so much that their internal quantum-mechanical components begin to "touch". Traditionally in astrophysics one talks of three main types of compact objects: white dwarfs, neutron stars and black holes.

We have recently discovered that an entirely new type of compact object, which we call "preon star", theoretically can exist if even smaller and more fundamental elementary particles than quarks and leptons, so-called preons or pre-quarks, exist. There are both physical and historical reasons why many more layers of smaller elementary particles should exist. If this proves true, then the possibility is high for objects even more compressed than neutron stars, *i.e.* there may be more destinies for matter before it, possibly, gets crushed into nothingness (a black hole).

If our new objects really exist, it would be something of a revolution in the understanding of matter and the fundamental physics laws. We have also shown that preon stars can be linked to several unsolved mysteries in astronomy: dark matter, cosmic rays with extremely high energies and possibly gamma-ray bursts. The results also show a connection between the size of elementary particles (*i.e.* the eventual substructure of quarks: preons) and the size and mass of the resulting compact objects. This means that it may be possible to do particle physics experiments, not by building more powerful and more expensive particle accelerators, but by doing, in this context, much less expensive astronomical observations. If the new elementary particles prove to be too small it may even be that we never, at least not in the foreseeable future, can create them directly in active particle physics experiments on Earth. However, it may be, relatively speaking, easy and inexpensive to detect them indirectly through passive astronomical observations.

Therefore, it is also very important to develop methods by which we can potentially observe preon stars. We are working to estimate the "signals" astronomers could expect to see because of these new compact objects: i) the gravitational waves that are markedly different from the ones expected from traditional astronomical objects, ii) gravitational lensing that also is unique (and actually may have already been discovered in existing data), iii) directional information from cosmic rays that can be created by rotating magnetic preon stars.

Theoretical issues that remain to be answered are, among other things: a) how and in what quantity preon stars can be formed in the early universe, b) if it is possible that preon stars may be a result of dying stars, c) whether and how the existence of preon stars change the expansion of the universe and gravitational structure formation and thus also the history and present age of the universe.

Why preons?

The most fundamental description of nature, which is also *tested* through experiments in particle physics, is the so-called standard model. There is the, supposedly, fundamental level of quarks and leptons, and the force particles mediating interactions between them. There are six known types of quarks (u, d, s, c, b, t) and also six different leptons (e, μ, τ, ν_e, ν_μ, ν_τ). A strange thing is that only the lightest quarks u and d are stable, the other quarks very quickly decay into these. The same is true for leptons, only the lightest are stable. Already from this fact one can begin to suspect that quarks are not as fundamental as once hoped. A truly fundamental particle should not be able to decay, right? Historically, unstable particles have eventually always been proven to be composed of more fundamental building blocks.

Preons is just such a proposed more basic level of building blocks. The first preon models were presented as early as the 1970s. The name "preon" was invented by Abdus Salam and Jogesh Pati in 1974. Salam later received the Nobel Prize, not for his work on preons but for his contribution to the electroweak theory, which today is part of the standard model.

In the early parts of this century, I myself was involved in the construction of a new, improved preon model, which can explain all the quarks and leptons in the standard model (not just those in the first "particle generation" as the first preon models did) in terms of only three fundamental and absolutely stable preons. Automatically, our model also explains a lot of other relationships, which in the standard model are completely unrelated and unexplained. Among other things, we obtained a relation between the Cabibbo angle and the Weinberg angle; in the standard model two completely independent parameters. It may thus reduce the number of free parameters in the theory, which in the standard model numbers over 20, most of which are related to explaining the different masses of particles, fixed through comparison with experiments. Our preon model also provides a number of experimentally testable predictions, such as new particles that possibly should be detectable using the particle accelerator LHC at CERN.

There is also a widely held view among the world's particle physicists that "something" novel should show up at an energy scale of about one TeV (Tera-electron volts, or 10^{12} electron volts, where 1 electron volt = 1.602×10^{-19} Joule is the natural energy unit for atomic physics). Some proposed candidates for this new physics are supersymmetric particles, extra dimensions, microscopic black holes, and *preons*. All agree, however, that the standard model cannot be the final theory; it contains just too many unexplained things and loose ends. One can also see that the theory is not really logically coherent, "seams" between its various components show too clearly.

Moreover, the whole electroweak part of the theory is more or less a mess. Preon models often have the advantage that they can clean up some of the mess pretty neatly.

One major unresolved question is how preon models can explain the relatively light masses of quarks and leptons. It has even been given a name, the "hierarchy problem". Binding forces contribute negative energies, so it could be natural to assume that the quark/lepton mass must be less than the sum of the masses of the preons that build them up, but all details have not yet been resolved satisfactorily. But to put it in perspective, for example string theory has exactly the same problem, but even worse; why is not every quark/lepton mass comparable to the Planck Mass (roughly that of a bacterium) the natural mass in string theory? According to string theory the only solution is that quarks and leptons are massless ... But they aren't! More precisely, it is the energy scale at which the new elementary particles show up that is relevant, and according to Heisenberg's uncertainty relation, the energy is inversely proportional to the particle 'size', *i.e.* the substructure length scale. The smaller the substructure you want to observe the higher energy you need.

Simply put, preons would be just the next step in the ladder of "fundamental" particles: atoms, atomic nuclei, protons/neutrons, quarks, preons, ... , ... the series of downward steps possibly ending with superstrings, but probably with something else still to be discovered. It is very naive to assume that there isn't something new between the scale of quarks & leptons and the (hypothetical) superstrings. After all, the relative difference in size between an atom and a superstring is as large as between the solar system and an atom, so there are probably many layers left to discover before we reach "rock bottom", if such even exists. I think it is extremely unlikely that we "hairless apes" would already have discovered all of nature's mysteries in the few hundred years' modern science have existed. In every epoch it has been believed that definitive answers have been found (or at least that they were just around the corner), but it has so far never been the case and the chance that we now might have found the ultimate "truth" is just as implausible now as then.

As always in science a theory is just a guess. One must then test it through experiments and/or observations. That is the "scientific method". What unsettles me a bit is that most of today's theoretical physicists are working on theories that are "untestable", at least for a very long time. It is reminiscent of the Dark Ages when "natural philosophers" in Europe debated how many angels could fit on the head of a pin, while Chinese astronomers made actual observations of real supernovae (exploding giant stars).

The real and practical difference between preons and strings is that preon theories can be tested today with the LHC and/or preon stars, string theory not in the foreseeable future.

Why preon stars?

Nature seems to work by the principle that all microscopic possibilities are realized. At the quantum level everything that is not strictly forbidden is compulsory! We have shown that a compact cosmic object consisting of preons is physically possible. This takes into account Einstein's general relativity, our most modern theory of gravitation, and the preon properties. If preon stars have high probability of being formed in abundance in the universe is still neither proved nor disproved. But because of the "compulsory" principle above, preon stars should exist if preons exist.

Preon Stars are not dependent on any particular theory of preons. The only requirement is that preons are fermions, a particle type defined by its quantum mechanical spin 1/2 (or 3/2, 5/2, etc.). Fermions obey the "Pauli exclusion principle" that forbids two fermions to be in exactly the same state. All currently known matter particles are fermions, so it seems natural that the next layer of matter should also consist of fermions.

The obvious novelty of preon stars in astrophysics is that they are a never previously proposed entirely new class of compact objects, which enlarges these to four from the traditionally "known" three; white dwarfs, neutron stars and black holes. We are just waiting for textbooks to be rewritten ... For theoretical particle physics, preon stars is a whole new way to test theories of what really are the fundamental constituents of nature. The reason that I myself have in some ways "defected" to astro-particle physics is that few really new results have arisen in experimental particle physics over the past 25 years. There are almost countless numbers of theories, most of which of course are wrong, but currently impossible to test; a sure sign that the theoretical "progress" is in the wrong direction. Physics is, after all, supposed to be about our description of nature, not some sort of quasi-philosophy. We try to base our own "crazy" ideas strictly in the framework of the scientific method. We recognize that our hypothesis about preon stars inevitably have to be backed up by experimental or observational evidence before it can reach widespread acceptance.

Preon Star properties

The traditionally recognized compact objects in the cosmos are, as previously stated, white dwarfs, neutron stars and black holes. White dwarfs are formed when gravity compresses a dead star, where energy production at the center has stopped and the normal pressure is lost, so that the atoms begin to "touch" one another. According to quantum physics (Pauli Exclusion Principle) there is in fact an effective "degeneracy pressure" between fermions even when the temperature approaches zero. The electrons in atoms are fermions. This gives the white dwarfs a size comparable to Earth. Above the so-called Chandrasekhar-limit (1.4 solar masses) the electron degeneracy pressure can no longer resist gravity. Put simply, gravity pushes the electrons into the protons and neutrons form. A neutron star is born when the neutrons are "touching" each other, because they too are fermions. A neutron star may be between 1.4 to 3 solar masses, and its size is only a few tens of kilometers. The density of a neutron star is comparable to that which would result if we compressed all the people on the earth into a volume as small as a sugar cube... If the stellar remnant is more massive than that, even neutron degeneracy pressure is not sufficient, and the star collapses indefinitely, into a black holes. Even so-called quark stars and hybrid stars behave almost as neutron stars, with the same limits for the masses. In quark stars the center of the star theoretically is a soup of free quarks, which gives a slightly smaller size and slightly different spin properties.

The above is based on the assumption that quarks and leptons have no internal structure, that they are not constructed from more fundamental constituents.

If preons exist the conclusions change completely.

Depending on the preon properties, a stable preon star will lie within a specific range of size and mass.

As particle physics has been tested up to an energy scale of about 1 TeV (1000 billion electron volts), any preons needs a higher energy than that. "High energy" is also, in quantum physics, translated to "short distance". This means that a preon has a "size" of no more than 10^{-19} meters, because matter has already been tested down to that level.

This results in preon stars that, at the most, may have a radius of about one meter, while the maximum mass is equal to one hundred Earths. Preon stars are therefore very much more compact than even neutron stars. This means that the new ladder of compact cosmic objects, in increasing order of density is: white dwarfs, neutron stars, preon stars, black holes.

The parameters that finally determine the size and mass of the preon star are i) the masses of the elementary particles themselves, the preons, ii) interactions, *i.e.* the forces between preons, iii) the gravity of the preon star. When the inward force (gravity) is exactly balanced by the outward (quantum degeneration) force, the preon star achieves static equilibrium. This is exactly the same principle as for white dwarfs and neutron stars, the only difference being the building blocks that give rise to degeneracy pressure.

Astrophysical puzzles that may be solved by preon stars

The most obvious role that preon stars can play in astronomy is that of "dark matter" in the universe. From various independent observations, we know that the matter content seems to be dominated by an exotic form of matter, which is never seen directly. These indications come from the rotation of spiral galaxies and the motion of galaxies in galaxy clusters. Both move too quickly to be held together only by the visible matter. As we also know that galaxies and galaxy clusters must have existed for billions of years and therefore must be stable, it means that we need a large amount of dark matter because they would otherwise be torn apart by their own rotation. The exotic matter is called "dark" because it does not radiate any light and barely interacts electromagnetically.

One can easily show that preon stars theoretically can be all the dark matter in the universe. Their smallness and compactness can explain why astronomers have not yet discovered them.

Very small preon stars can probably be produced in sufficient quantities from the density fluctuations in the early universe. This also has the advantage that it avoids a limit that exists for how much normal matter there may be in the universe. The Big Bang theory automatically provides constraints for how much normal matter can be formed, but because preon stars can be created much earlier, and also is not normal matter, they are not affected at all by this theoretical limit.

Yet another astronomical mystery is the origin of the most energetic cosmic rays. There is a theoretical limit, known as the GZK-limit, on how high the energy of a cosmic ray can be. If they exceed a certain energy the universe is no longer transparent to cosmic rays. The universe is not empty, but is everywhere filled with approximately 400 photons per cubic centimeter from the cosmic microwave background, the cooled down remnant from the Big Bang. No known objects that can accelerate cosmic rays to the observed immense energies are within, in cosmological terms, the short distance required. Preon stars fits perfectly, however, as the origin of these high-energy cosmic rays, because they may be "nearby" and at the same time can have huge magnetic fields and rotational speeds sufficient to accelerate such protons to the highest observed energies (and beyond).

Methods for detecting preon stars

Since preon stars are so small, as footballs or less, the chance to observe them directly with telescopes is near impossible. One method of detecting them is instead based on gravitational lensing. In Einstein's general relativity, gravity is not a force but a curvature of space-time. This means that an accumulation of matter can act much like an ordinary optical lens, which has also been observed several times. Despite their small sizes preon stars still have sufficient mass to affect the light from distant light sources. Preon stars are however far too small to give the normal characteristic of a gravitational lens; short-term increase of a distant object's brightness. Instead one gets a diffraction effect similar to that of a regular lattice. For preon stars this effect lies in the X-ray and gamma parts of the light spectrum.

Therefore could gamma-ray bursts be excellent light sources for observing this type of phenomenon. Gamma-Ray Bursts are huge, short outbreaks of high-energy radiation that are believed to be created in massive exploding stars and collisions of neutron stars. Since GRBs radiate most of their energy in the short gamma-and X-rays, preon stars can lens this radiation. We have also pointed out that the gamma-ray burst observed by the Japanese satellite Ginga on February 5, 1988 appears to show just such a lensing by a preon star. The spectrum from Ginga showed two dips in intensity right at the wavelengths that would result if an intermediate preon star acted as a gravitational lens. Astronomers have previously discovered a few candidates of just this kind, which gives an indirect indication that preon stars exist. However, more detailed observation and analysis must be done before firm conclusions can be drawn.

Yet another method to indirectly "see" preon stars is to study cosmic rays at extremely high energies, e.g. through the Auger Observatory. Because these cosmic rays have such high energy they are not influenced by the Milky Way's magnetic field, and may therefore give direct information on where a preon star is, *i.e.* its direction on the sky. This can then be supplemented with follow-up gravitational lens observations, etc.

A third method we proposed is based on an additional relativistic phenomenon: gravitational waves. These waves are "ripples" in space-time geometry caused by the acceleration of massive objects. Two preon stars that whirl around their common center of mass may be possible to detect with future experiments if they are within a few thousand light years from Earth.
A binary pair of preon stars in a relatively wide path would generate gravitational waves with a frequency of a few thousand cycles per second. The proposed (underground) Observatory EURO, which will be completed within the next ten years, would have enough sensitivity to see them if they are there.

Because of their small sizes, some preon stars swirl around in an orbit less than one millimeter. Such a system would generate gravitational radiation with millions of cycles per second. The University of Birmingham has built a prototype that uses resonant microwaves to detect such gravitational radiation. Although the current prototype does not have enough sensitivity to see binary preon stars, the next generation of instruments may. Detection of extremely high frequency gravitational radiation would provide detailed information on the densities and masses of the sources and be able to demonstrate the existence of preon stars.

Future challenges

A key unanswered question is: How are preon stars made? There are probably two main ways by which they can be created: either by density fluctuations in the early Universe or by dying stars in the later universe (or both). That there must have been variations in the density of the early universe we know already, otherwise no objects at all would have formed. The question is whether they have the right properties to create preon stars in such quantities that they may constitute the dark matter.

Another important thing is to examine whether it is possible for a star of several solar masses to expel the bulk of its mass in its "death rattle", and thus be able to form one or several preon stars instead of a black hole. Today there are no easy answers to this. It is intuitively difficult to understand how preon stars, with masses comparable to Earth's, can be created from super-massive stars. But it has never been proven impossible. In fact, very little is known about the details of how giant stars explode. For example is it almost impossible to get a model supernova to explode in computer simulations. Realistic simulations would include non-symmetrical collapse, strong magnetic fields, turbulence, neutrinos, neutrino-like radiation at the preon level, gravity waves, and so on. Then it may turn out that i) almost all the mass/energy is radiated away and leave a light preon star, ii) many preon stars are formed like "pearls on a string", especially if the interaction between preons behave similarly to QCD, the theory of how quarks interact through the exchange of gluons. In QCD it is well known that composite particles are created as "pearls" along the relativistic jets.

In fact, the actual time-dependent formation of a dynamic compact object is very poorly known. Investigations of the properties of static compact objects, once they are formed, are in comparison almost infinitely easier. So there is yet nothing to say that a preon star of the order of an earth mass cannot *in principle* be formed by a dying star of the order of ten solar masses.

One of the most important challenges is to develop exact theoretical signatures that can be compared with astronomical observations. We have already taken the first steps in that direction, but more precise and exhaustive analysis is needed.
Indirect indications of preon stars can also be generated by simulation of dark matter in galaxies, clusters of galaxies and in the whole universe, especially as concerns structure formation which can then be compared with future observations.

The most direct method would of course be if preons were detected in particle accelerators on Earth. However, this will only work if preons are relatively light. For very massive preons there is no chance of seeing them directly in accelerators in the foreseeable future. If so, one can instead turn the tables around and use indirect astronomical observations to deduce the structure of matter on the fundamental level. Existing astronomical observational methods can explore preon "sizes" that might not be directly observable in accelerators for hundreds of years, if ever. The higher the energies, *i.e.*, the smaller the sizes we want to study – the particle physics research frontier - all the more reason there is to look outwards and upwards instead of inwards and downwards... Actually, the working title for one of our scientific articles on preon stars originally was "Are we looking in the right direction for fundamental constituents?"

In summary:
1) We have discovered the theoretical possibility of an entirely new compact cosmic object, the Preon Star. That is, we have expanded the number of classes from three (white dwarfs, neutron stars, black holes) to four (white dwarfs, neutron stars, preon stars, black holes).
2) We have proposed several ways through which astronomers can observe preon stars, if they exist.
3) And most important to me as an "old" particle physicists, we have shown how one indirectly, "cheaply & easily", can observe the substructure (preons) of quarks and leptons (if it exists) by astronomical observations. Even for extreme energies that the LHC will not have access to (or any other particle accelerator in the foreseeable future).

Related scientific articles:

"The observational legacy of preon stars - probing new physics beyond the LHC" Physical Review D76:125006 (2007)
http://arxiv.org/abs/astro-ph/0701768

"Preon stars: a new class of cosmic compact objects" Physics Letters B616 (2005) pp 1-7 http://arxiv.org/abs/astro-ph/0410417

"Preon Trinity - A Schematic Model of Leptons, Quarks and Heavy Vector Bosons" Europhysics Letters 57 (2002) pp 188-194
http://arxiv.org/abs/hep-ph/0208135

MODERN PHYSICS IN 15 MINUTES

ABOUT THE AUTHOR

Johan Hansson is Professor of Theoretical Physics, presently at Luleå University of Technology in Sweden. Some institutions at which he has previously been active include UC Berkeley, the University of Torino and Richmond University. Prof. Hansson is, among many other things, the "inventor" of *Preon Stars* and *Neutromagnets* as covered in Nature, Physical Review Focus, New Scientist, Physics World, MIT Technology Review, etc.

www.ingramcontent.com/pod-product-compliance
Lightning Source LLC
Chambersburg PA
CBHW061513180526
45171CB00001B/159